Rebecca Schwarzlose is a neuro
St. Louis. She holds a PhD in Ne
as chief editor of the scholarly
She lives in St. Louis.

'Broad in scope, and beautifully written'
Joseph LeDoux, author of *The Deep History of Ourselves*

'Rebecca Schwarzlose is a neuroscientist with a novelist's literary flair. *Brainscapes* is a profoundly illuminating account of how the brain works – and of how the maps within our heads determine what we see, recognise, remember and feel … a complete inspiration'
Cass R. Sunstein, author of *Too Much Information*

'This book is the Lonely Planet travel guide to the brain. With humility, humour and the familiarity of a local, Rebecca Schwarzlose takes you by the hand and shows you around some of the strangest landscapes of the cortex. In *Brainscapes*, the brain becomes an open atlas'
Stanislas Dehaene, author of *Consciousness and the Brain*

'Clear, often vivid history'
Nature

'Enlightening and ambitious … a book that travels into rich terrain, charted by a smart and eager tour guide'
The New York Times

'In lively prose, Schwarzlose introduces you to your inner cartographer: a complex brain that continuously constructs shifting maps of the world, charted from the perspective of your own body. These maps are not just created by you – they are you. They conjure what you feel, what you remember, and what you do'
Lisa Feldman Barrett, author of *Seven and a Half Lessons About the Brain*

'*Brainscapes* will change how you think about the brain and how you understand your own mind. A fascinating and original exploration of the physical principles that enable you to do all that you do, and be who you are'
Tali Sharot, author of *The Influential Mind*

'A beautiful book about one of the most fundamental properties of the brain – its ability as a mapmaker. The meat in our heads organises and controls everything we do, from perception to emotion, action to cognition, by mapping complex information into simple spaces. *Brainscapes* explains that deep truth in clear, compelling language. It's a fascinating, well-told story'
Michael Graziano, author of *Rethinking Consciousness*

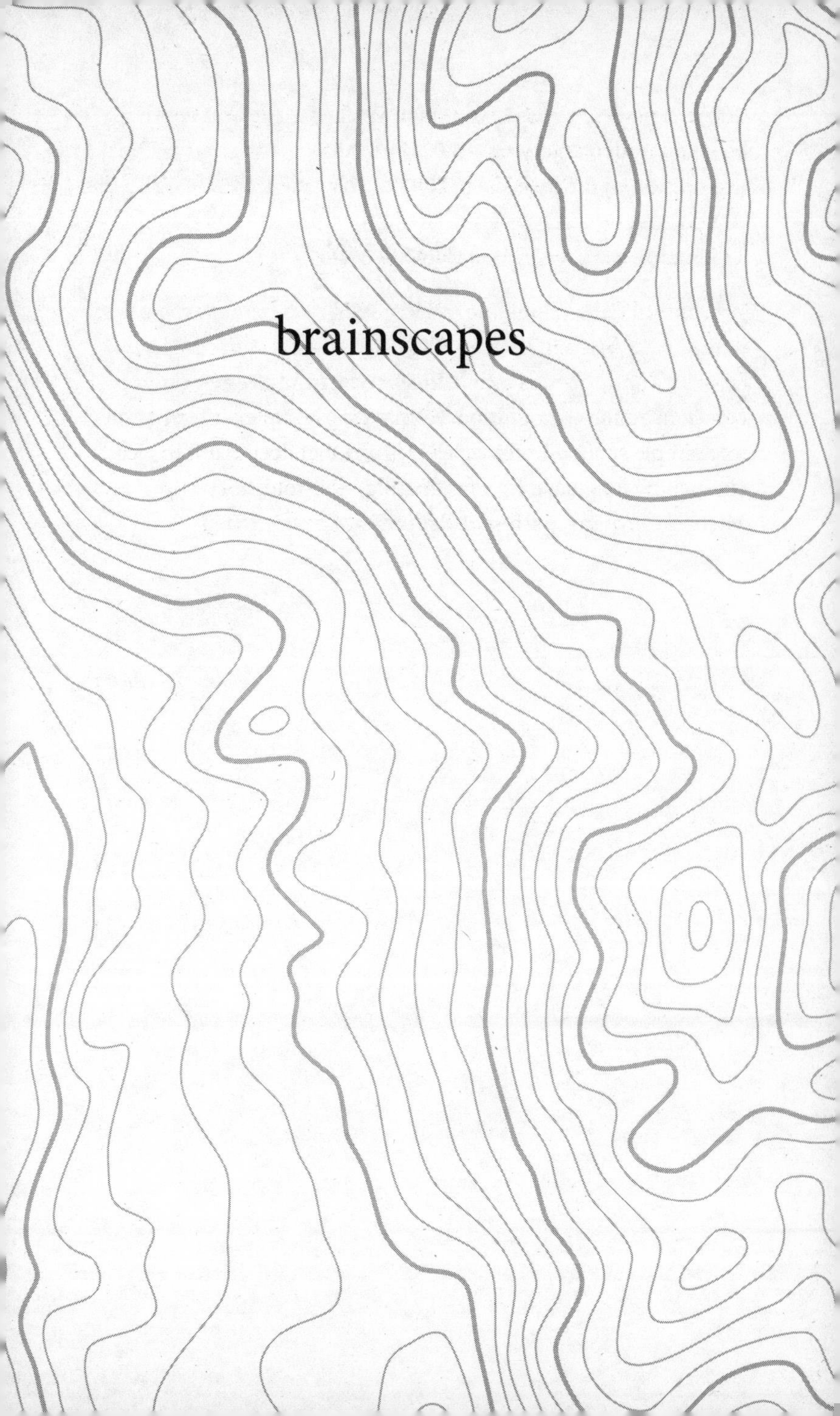

brainscapes

brainscapes

AN ATLAS OF YOUR LIFE ON EARTH

Rebecca
Schwarzlose

P
PROFILE BOOKS

This paperback edition first published in 2022

First published in Great Britain in 2021 by
Profile Books Ltd
29 Cloth Fair
London
EC1A 7JQ

www.profilebooks.co.uk

First published in the United States of America
by Houghton Mifflin Harcourt

Book design by Chrissy Kurpeski

1 3 5 7 9 10 8 6 4 2

Printed and bound in Great Britain by
CPI Group (UK) Ltd, Croydon, CR0 4YY

A CIP catalogue record for this book is
available from the British Library.

ISBN 978 1 78816 053 7
eISBN 978 1 78283 449 6

This book is dedicated to my mother,
Sally Frye Schwarzlose,
whose inspiration and support
made it possible

Contents

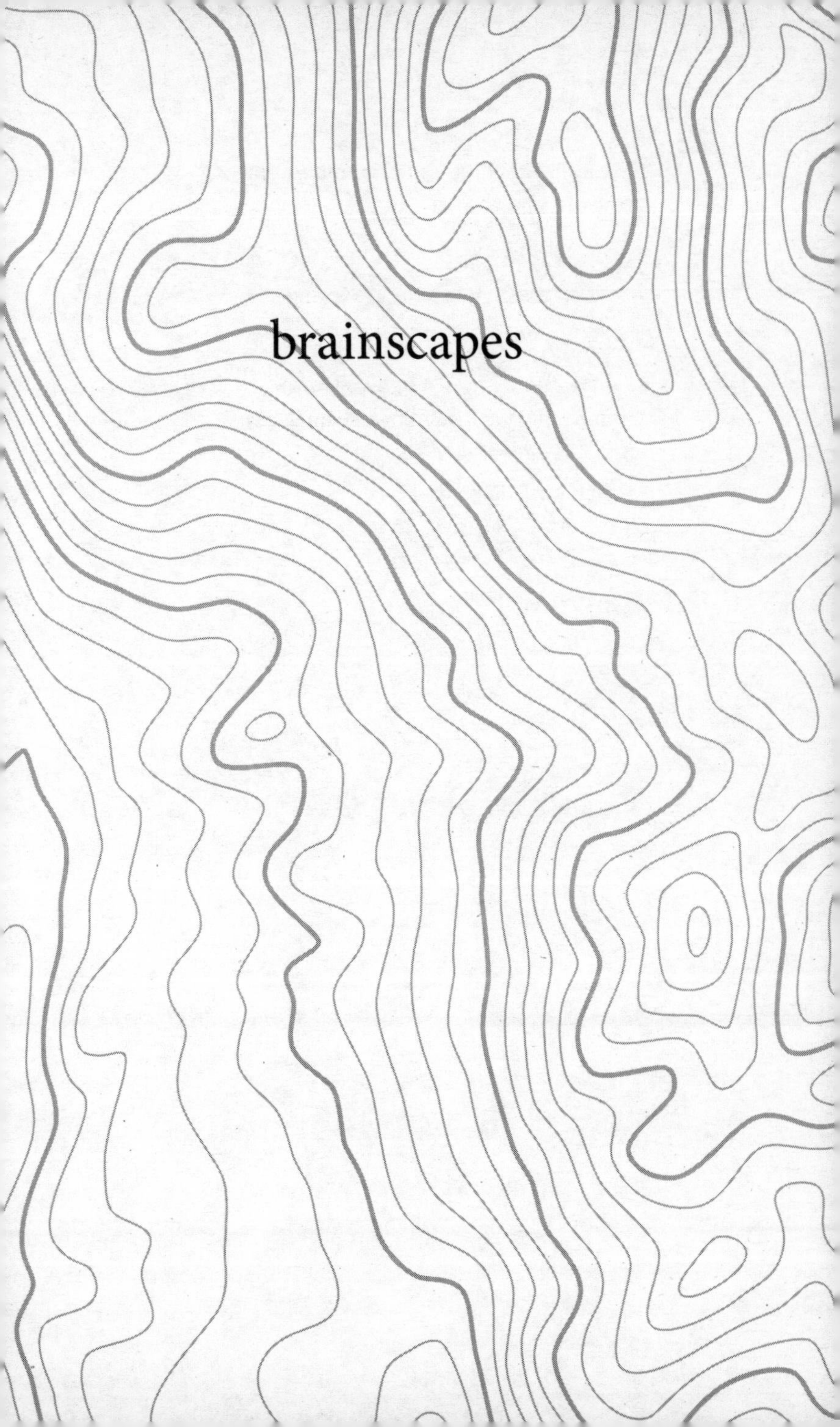

brainscapes

Introduction

LET ME BE CLEAR. This book is about real maps in your brain. I am not being metaphorical or using artistic license; there are actual maps in your brain. And not just one or two maps. Your brain is a teeming atlas of them.

Before I explain what I mean by a brain map, here's what I don't mean. Scientists and science writers often describe neuroscience research as "brain mapping" in news articles, popular books, and lectures. By this they mean the scientific effort to discover, for example, which chunks of the brain make you feel sleepy or help you select one dish from a crowded restaurant menu. This so-called brain mapping doesn't create a map so much as an inventory of known brain regions—a list that resembles a mechanic's list of automotive parts. Each part, from the crankshaft to the muffler, has a specific function and belongs in a specific place in the car. Likewise, this brain inventory lists brain chunks, their locations within the brain, and their suspected purpose. But once neuroscientists discover an area of the brain to add to the inventory, they inevitably zoom in and investigate it closely. When they do, they often find that this chunk of brain has its own internal organization—an astonishing layout that forms an actual map.

Simply put, a map is a spatial representation of something else. When you think of maps, you probably think of geographic maps, which represent locations on the surface of the earth. But you could just as well map locations on the surface of the moon or the arrangements of stars in distant galaxies. In fact, maps are not limited to representing physical locations. The maps in your brain chart your body, senses, movements, and crucial sources of information in your world.

To grasp the beauty of brain maps, you need only think of their coordinates. Geographic maps have spatial coordinates, or units of distance such as kilometers or degrees of longitude and latitude. A point on a geographic map gives you a location on the surface of the earth. But a point on a brain map may give you a flash, a flavor, a note, a twitch, or a tingle. Your brain maps of vision represent the droplets of light that strike the delicate sheet of cells at the back of your eyes. Your maps of hearing represent frequencies of sound impinging on the sensitive coils buried deep within your ears. Your touch maps have skin coordinates and represent each instance of pressure, warmth, and pain along the many plains and valleys of your skin. Your brain contains spatial maps, which help you interact with the world, and movement maps, which plan what you are about to do: the next blink, swallow, or step. Literally, these are maps of *you*—your body, from your eyeballs to the bottoms of your feet—and what you do, what you need, and how you interact with the spaces around you. Your brain is brimming with remarkable maps that represent these facets of your world and more.

But brain maps aren't just remarkable. They are also *important*. They offer deep insights into us and why we remember, imagine, learn, and think the way we do. They allow us to peer into the minds of other people, even other creatures, and infer how their perceptions differ from our own. They expose how children learn from and adapt to their surroundings, sometimes with lifelong repercussions for health and well-being. And, increasingly, physicians, scientists, and corporations are using brain maps to develop technologies that bridge the divide between computers and brains. In short, brain maps hold profound psychological, philosophical, societal, and technological importance. They are a key to understanding ourselves in the present and to deciding how brain technologies will shape our societies into the future.

In the process of writing this book, I found that several themes emerged in the story of brain maps. The first is their universality. From fruit flies to catfish to elephants, the brains of creatures great and small contain maps. And not just one or two! From the front to the back of the brain, from left to right and top to bottom, maps abound. It is simply what brains on earth do: they make maps. And in many cases, they make similar maps. You and I have more or less the same types of brain maps, and we share many of them with species cloaked in scales, feathers, or fur.

The second theme is the uniqueness of brain maps—a concept that might seem to contradict their universality. But it does nothing of the sort. Universality and uniqueness are close companions throughout the biological world. For instance, fingerprints are universal to humans—we all have them—and yet the idiosyncrasies of individual fingerprints make each one unique. The same is true for brain maps, except that their idiosyncrasies matter. They affect a person's or a creature's capacity to perceive, attend, remember, recognize, and react to the wider world. As such, brain maps offer clues about how perception and even certain types of aptitude, including reading ability, visualization, and fine motor ability, differ from one person to the next.

A third theme is that brain maps are born out of necessity. They are not an ornamental flourish in nature's grand design. Without them, perception as you know it would not be possible. There are tremendous physical challenges to creating, fueling, and housing a brain capable of seeing and feeling worth a damn. In fact, making brain maps is only the first step in surmounting these obstacles. To truly overcome them, your brain maps cannot faithfully represent the world around you. Instead, they are dramatically warped to save energy and space, and these distortions, in turn, shape what you can see and feel. This surprising fact lies at the heart of this book and its title. As you will see, brainscapes are the distorted representations of reality that occupy your brain maps and dictate what you perceive.

Fourth, the story of brain maps is one of tough trade-offs, not superiority. If you came to these pages hoping to learn why some people are smarter than others or why humans outshine other creatures on earth, I recommend that you look elsewhere. Brain maps reveal a different truth: one of limited resources and penurious investment. When a species invests and excels in particular abilities such as high-definition vision, manual dexterity, or ultrasonic hearing, other capacities are by necessity ceded and sacrificed. Thus the brain maps found within a given creature—including you—reveal the perceptual and mental abilities crucial to its survival. A brain map can't be judged superior or inferior based on its intrinsic qualities; its value can only be determined in the context of a creature's environment and moment-to-moment needs for survival.

Brain maps also give organisms the chance to adapt. Throughout life, these maps retain some capacity to learn from experience and reshape themselves in response to new environments and needs. The brains of infants and young children are, however, especially pliable, and so experiences that occur during key weeks or months of childhood can dramatically impact how the maps will be laid out for a lifetime. This process of learning from their environment affords children the chance to adapt to it, in some cases by altering their brain maps to a dazzling degree. Typically, this process benefits children and helps them to thrive throughout life. But when a child's early environment is disrupted by transient yet extreme experiences such as hospitalization, deprivation, or correctable visual impairments, this kind of learning can have the opposite effect. In essence, brain maps reveal how children can be particularly resilient when facing changes in their environment and yet also particularly susceptible to suffering enduring effects as a result of those experiences.

Finally, a brain map provides an ideal conduit for eavesdropping on the living, thinking brain—and even directly manipulating it. In the late nineteenth century, knowledge about a brain map enabled physicians to begin performing successful targeted neurosurgery. In the present day, knowledge of additional brain maps allows scientists to detect when people who appear to be in a vegetative state are, in fact, fully conscious and aware of their surroundings. It permits some forms of mind reading—not just deciphering what a person is seeing or feeling at a given moment, but also what they are remembering, imagining, or dreaming about. It has spurred advances in neuroprosthetics aimed at restoring movement to patients suffering from paralysis and vision to people who are blind. And recently, commercial attempts to directly integrate computers and the human brain—a prospect that has opened the wallets of private investors and corporations alike—are likewise based on information about brain maps.

As these themes collectively suggest, the impact of brain maps on thought, health, and technology is profound and far-reaching. They matter not just to scientists but to every person and creature on earth. They matter to you, though you may never have heard of them, and they offer answers to questions so fundamental to daily life that you may never have thought to ask them: *Why does remembering an event feel like reliving it? Why do I use my hands to feel things? Why can't I imagine sensations beyond my five senses? Why are children better than adults at learning new skills like playing an instrument? Why can't I pay attention to everything around me all at once? Why are computers so good at predicting the next thing I might type but so bad at understanding what I actually mean?* I wrote this book not just to answer such questions but also to revel in asking them.

Brainscapes is not a how-to book. It will not improve your golf game or your love life. It offers no treatments for illness nor tips for financial success. These pages offer something else entirely: a chance to view your life from a different perspective. So much about how you feel and think may seem obvious. But beneath the ordinary, there is plenty to marvel at. Brain maps shape your experience, day in and day out. They tell a grand story—one that is both universal and deeply personal. They reveal surprising truths about our place in the world and about the world's place within us. They raise important questions about what is real, what is fair, and what is private. And they demonstrate how dire necessity can give rise to something surprisingly beautiful and even poetic.

This book introduces only a handful of brain maps, each one vital for perceiving the world and acting upon it. It reveals what brain maps are, why they exist, and how your body teaches your brain which maps to form. It shows how brain maps serve as the loom upon which you weave the threads of perception, attention, imagination, and dreams. It describes how they shape your thoughts and abilities, how they empower you yet at the same time hold you back.

The brain is often described as immensely complex. Writers and scientists sometimes compare it to the stars in the heavens in order to illustrate the sheer number of cells in the brain and the stunning tally of connections between them. Although such comparisons inspire a justified awe, they also portray the brain as distant—unreachable and unknowable. But to consider the brain through its maps is a powerful means of understanding it as a close ally. Your brain is intimately attuned to your needs and your self—the layout of your body, the sensitivities of your flesh, your abilities and limitations, the space that surrounds you, and the tools and fuel you rely on to survive.

Yes, your brain is complex. But it is not out of reach, far off in the heavens. It is down in the mud with you. And that is a truly wonderful thing.

1

An Atlas of You: What Is a Brain Map?

As with so many scientific discoveries, it took both serendipity and suffering to reveal vision's hidden maps within the human brain. The first of these maps was unearthed because of two things: bullets and blindness. The story of how they intersected illustrates what a brain map actually is and why it matters.

The year was 1904. Two empires, Japan and Russia, were locked in a war over territory and naval dominance. As casualties on both sides mounted, the Japanese authorities noticed something strange: a surprising number of their wounded soldiers had become fully or partially blind. Many of these cases, as might be expected, involved injuries to the eye. But nearly a quarter were due to brain damage—a proportion far greater than that of previous wars.

This shift reflected a change in weapons. The Russians were using a new type of gun: a high-velocity rifle called the Mosin-Nagant Model 91. Its bullets were 7.6 mm in diameter, smaller than bullets fitted to previous rifle models. Clocking speeds of 620 meters per second out of the barrel, they were faster as well. These changes enabled the Russians to fire bullets that traveled farther yet still hit their mark. They also changed the types of injuries these weapons caused. Previously, gunshot wounds to the head often shattered the skull or sent shock waves through tissue, creating large cavities in the brain. Not so with the Mosin-Nagant Model 91. The new bullets sliced clean through both brain and bone, leaving small, neatly circumscribed holes like perfect rounds in a paper target. As a result, more soldiers survived gunshot wounds to the head, and those who did were left with fewer mental and perceptual problems than those with head wounds in previous wars.

The Japanese soldiers who were shot in a certain part of the head reported visual problems. They experienced a hole in vision—a blind patch called a scotoma—and it traveled with their gaze as they looked from one place to the next. Damage to the eyeball can cause a person to experience a scotoma. Yet the wounds that caused these particular scotomas were in an odd location—the back of the head, far from the eyes.

These holes in vision differed, both in size and location, from one patient to the next. One man might have a scotoma in the far right-hand corner of his vision, so that wherever he looked, objects on the right would be obscured. Another man might be unable to see the very thing he fixed his gaze upon; whether he tried to look directly at a written word or at the face of his wife, the blind patch would eclipse it.

These scotomas marred a patient's vision whether he tried to see with only his left eye, only his right, or both eyes at once. It became clear that the scotomas would handicap the soldiers for life. To compensate for these injuries and lost future income, the Japanese government planned to award the men larger pensions—a complex undertaking, in the bureaucracy of the empire. In order to determine the size of each soldier's pension, the government needed documentation of the location and size of his scotoma.

The somber task of gathering that information would fall to Dr. Inouye Tatsuji, a young ophthalmologist who was just finishing his medical studies when the war broke out. Inouye's work entailed performing visual field tests on the wounded soldiers. A visual field is the expanse that a person can see in all directions without moving the eyes. When you fix your gaze on a point in space, you can still see a certain distance above, below, and to either side of it. If your vision is normal, you will see best at your center of gaze, which corresponds to the point in space where you are looking. Yet you also collect visual information far from your center of gaze, such as when you notice someone "out of the corner of your eye." Vision scientists and ophthalmologists call these regions on the outskirts of the visual field the visual periphery. Your visual field is the sum of your vision at your center of gaze and in your visual periphery. Ophthalmologists can test a patient's visual fields in many ways, from flashing lights at different places in the visual field to simply holding up fingers in various parts of the visual field and asking the patient to count them. When a patient fails to see anything in a portion of that field (say, the lower left-hand portion), they might have a scotoma.

During his time as an army physician, Inouye performed countless visual field tests on patients with gunshot wounds or other traumas, and he compiled records of everything he found. Although his government simply needed the records to calculate pensions, the young doctor realized that they might also hold the key to resolving a long-standing scientific mystery: where was the seat of visual perception in the human brain, and how exactly was vision represented there? For decades, scientists had labored to answer these questions. Their work hinted at the existence of a map for visual information in the brain, but its exact location and layout remained unknown.

A map is a spatial representation of something else. Therefore, a brain map is a spatial representation *in the brain* of something else. Brains have sizes and shapes, fronts and backs, tops and bottoms. So having something *spatial* in the brain, something that takes up space in its surfaces, bundles, and folds, is easy enough to accept. But what does it mean for your brain to *represent* something about the world, whether that something is the sight of a painting or the sound of a siren?

The answer to that question begins not in the brain, but elsewhere in the body. Or rather, the key places where the body meets the outside world. Your skin is the largest and most obvious of these, but there is also the retina that lines the back of each of your eyeballs. There is the cochlea nestled deep inside each of your ears. There are the delicate tissues lining the inside of your two nostrils, and there are the moist surfaces of your tongue and mouth. And that's more or less it. These surfaces are the conduits through which you collect information about the outside world. Each of these surfaces is lined with cells called sensory receptors, which detect information from the external world and translate it into internal messages that will be sent to the brain. These receptors are unimaginably precious. Life without them would be a life unmoored from your surroundings.

Take a closer look at one of the features that anchors you to reality: your skin. It forms a continuous surface. If you focus on any spot of skin on your body, you will likely find more skin to the left, to the right, above and below. Sure, there are interruptions, such as the eyes, the mouth, and the nostrils. But the skin continues around them, just as the shore continues around a lake. The skin on your foot is next to your ankle skin, which is next to your shin skin. In other words, your skin has features that are consistently arranged across a continuous surface. That means your skin, like the surface of the earth or the moon, has a topography, or landscape. Imagine some microbes living on your skin. If colonies of microbes could communicate and explore, they might chart out the landscapes (or perhaps skinscapes) of your body and travel around, guided by its landmarks. Want to get to the armpit? Hang a right at the belly button and a left at that oddly shaped mole.

Although your skin is continuous, its touch receptors are not. Many thousands of receptors are embedded in your skin. Some detect trauma and signal pain, whereas others register pressure, vibration, or heat. Consider the set of receptors that specifically registers pressure and vibrations on your skin. Thanks to them, you are able to feel a poke and distinguish a smooth surface from one that is rough. Each of these receptors works its magic at only one spot on your skin. A receptor on your right kneecap is tasked with detecting and broadcasting touch on one patch of that kneecap. That is all. Think of it as something like a reclusive landowner hunkered down with a shotgun: *The rest of the world may do what it pleases, but if someone sets foot on my land, there'll be hell to pay!*

For that little kneecap receptor, all that matters is its kneecap territory. Scientists call this its receptive field; it is the field, or zone, from which that cell receives information. Pressure within its receptive field triggers a response, or message, from the cell. *Something is happening!* Pressure outside that receptive field triggers nothing. Like landowners, some receptors are responsible for bigger areas than others, but all receptors are restricted to their local plots of skin. Don't ask a receptor on your kneecap what is happening on your back. It couldn't tell the difference between a luxurious back massage, a slap on the back, or no touch at all. Each sensory receptor simply tells the story of its little patch of skin and sends that on to the brain.

This is where representation begins. The signals sent from an individual receptor, say, one embedded in the skin of your right knee, represent compression of that skin. So if I want to know whether your knee is being pressed, I don't have to inspect your knee. Instead, I could listen in on the signals sent from your skin to your brain. The signal from this receptor tells me all I need to know about that single patch of skin. The signal *represents* the physical forces acting upon that one part of your body.

Imagine that we follow the signal sent by the receptor on your knee and travel with that signal into the brain, where it reaches a type of brain cell called a neuron. Not just any neuron, mind you; this is no willy-nilly pairing of sender and receiver, like some cartoon of an old-fashioned switchboard. The precious signal will land upon just the right neuron: a specialist that gathers signals about touch but not sight, taste, smell, or hearing. Not just that, but a *super*-specialist that gathers signals about touch only on and around your knee, not your elbow or your face. Even though this neuron is in your brain rather than on your skin, it has a receptive field on your body: a plot of skin on your knee. That is all it knows; it receives information about that solitary site.

Likewise, when this neuron is ready to send a signal of its own to other parts of the brain, it can speak only to what it knows: information about touch on your knee. Even though this neuron is in your head and not on your kneecap, its signals will represent touch information from your knee. When this cell sends a signal to other places in the brain, this signal *means* something. It represents what has happened on that specific patch of skin. This is the central idea behind representation in the brain and a necessary ingredient of brain maps, not to mention pretty much everything the brain achieves. If brains did not create such representations, we would be done for. Brains allow us to gather clues from our sensory receptors and send instructions to our muscles *only* by representing what receptors sense and how muscles move.

Because of such representation, neuroscientists can now detect what a person is feeling on their skin simply by observing activity in their brain. Conversely, they can manipulate what a person thinks they feel on their skin by directly mucking around with their brain. That is the powerful thing about representation: once you know how it works, you can use it to listen in on signals or even change them.

Representation works nearly the same way for vision as it does for touch. Vision begins at the back of the eyeballs. When tiny packets of light, or photons, enter your eyes, they travel through your eyeballs and land on the delicate tissues of your retina. Embedded in each retina are many millions of sensory receptors that detect photons.

Just as skin is continuous, the retina that lines the back of the eyeball is a continuous sheet. And just as your skin has a topography, so too do your retinas. For instance, each retina has a conspicuous pit called the fovea; when you look directly at something—say, a stop sign—light that bounces off that sign and enters your eye will fall upon a sensory receptor in your fovea. Because light travels in a straight line and because the receptors in your eye are anchored in place, a receptor in your fovea will only ever detect and represent light coming in from your center of gaze, the area you are currently looking at. Likewise, a receptor far away from your fovea will only ever detect and represent light coming from elsewhere, a specific re-

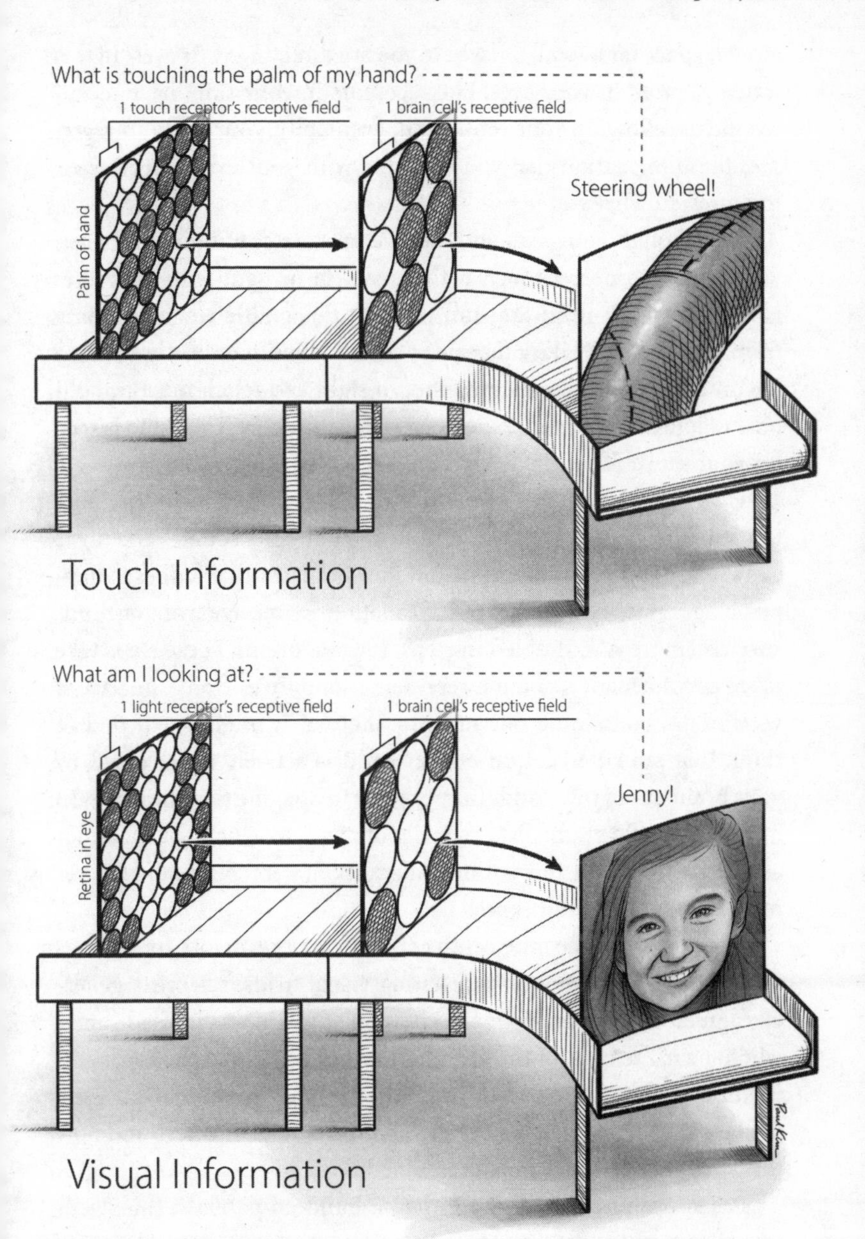

FIGURE 1. A schematic illustration of how receptive fields support neural representations in touch (above) and vision (below). *Paul Kim*

gion of space far away from where you are directing your eyes. In this way, receptors in your eyes, like receptors in your skin, have receptive fields, allowing your retina and, eventually, your brain to *represent* the information that you collected with your eyes—in essence, to represent what you see.

So although your skin and your retinas are continuous surfaces, your ability to detect touch with your skin or light with your eyes is not. It is built from the sum of all of those little signals coming from tiny patches. Like a mosaic of colored tiles that together form a meaningful image, your experience of sight and touch must be cobbled together from fragmented bits of information. The same is true for your sense of hearing.

How are these shards of perception integrated to form the more unified experiences we have of touch, sight, and sound? Scientists can't fully answer this question, but they do know that it doesn't happen all at once. The mosaic of information we receive from our sensory receptors is assembled in steps. (See Figure 1.) These steps take place as information about a representation moves from one part of your brain, containing one map, to another. It might seem odd to think that your perception of the world is actively constructed by your brain out of thousands of specks, or to imagine that these specks merge and meld gradually, in a series of maps, to generate experience as you know it. But this is the odd reality of perception and the remarkable nature of our senses.

By the close of the nineteenth century, not long before Inouye was examining bullet wounds and testing visual fields, scientists generally agreed that the brain's representation of vision was housed somewhere near the back of the brain. They also knew that the representation was spatial and that its layout in the brain somehow mirrored the layout of light entering the eye. Yet the details of where and how this strange map was laid out remained unclear.

Based on his study of more than a hundred patients, the Swedish neuropathologist Salomon Henschen had correctly identified the specific place at the back of the brain that housed the visual representation. He even proposed a theory for how the map was laid out in this area, but his description would prove wrong. The damage in his patients' brains was too widespread to make accurate observation possible. A little more than a decade later, young Inouye would triumph where Henschen had failed, and he would owe much of this success to the brutal efficiency of the Russians' new guns. The clean, neatly circumscribed bullet holes made by these guns, and the smaller scotomas that they created, made it possible to link bullet holes to blindness and, in doing so, to uncover the hidden map of vision in the soldiers' brains.

Inouye recognized the importance of precision. If he hoped to accurately chart the seat of vision in the brain, he would need painstaking measurements of both the scotoma and the location of the bul-

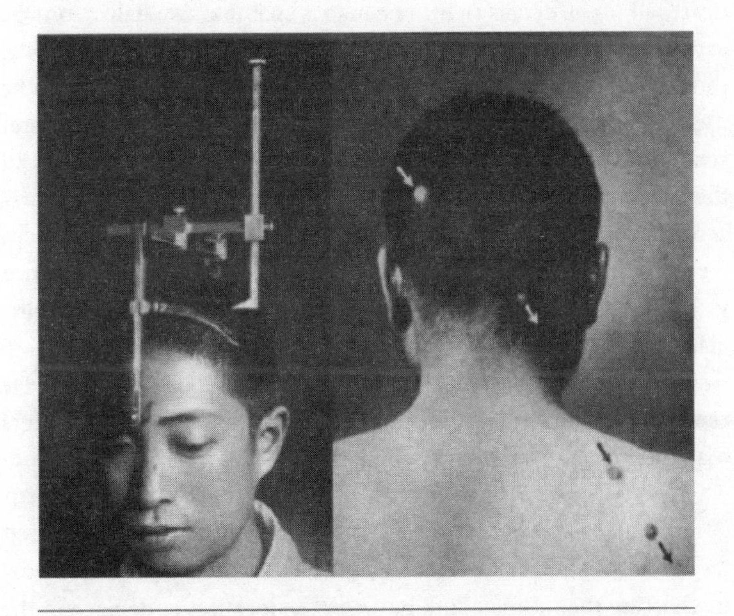

FIGURE 2. Photographs of a soldier in Inouye's study, showing the cranio-coordinometer in use (left) and the path of the bullet through the soldier's body (right). ***From*** **Die Sehstörungen bei Schussverletzungen der kortikalen Sehsphäre** ***(Visual disturbances following gunshot wounds of the cortical visual area) by Inouye Tatsuji (Leipzig: W. Engelmann, 1909).***

let hole in each soldier's head. Visual field tests were already accurate and commonly used, but Inouye had to conceive his own method for measuring and comparing brain damage from one soldier to the next. He invented an instrument called a cranio-coordinometer, which was essentially a series of rulers attached to one another in three dimensions by means of adjustable screws. (See Figure 2.) The instrument could rest on a person's head like a cap, which allowed Inouye to accurately measure the dimensions of different heads. He extrapolated each bullet's three-dimensional course through the brain and compared this with the location and extent of the blindness each patient experienced.

In 1909, Inouye published a monograph presenting his findings from twenty-nine soldiers. His report detailed an actual map of visual space in the human brain. And unlike Henschen, Inouye got nearly all of the details right. The map is split into two halves, one on each side of the brain. Both halves are tucked away at the very back of the brain, in an area that we now call the primary visual cortex, or its nickname: V1. Compared to the visual field it is based on, the representation is upside-down: representations of grass and soil sit above those of sky and clouds in the actual tissues of the brain. The representation is also left-right flipped, so that the right visual field is represented in the left side of the brain, and vice versa. Furthermore, this visual representation is profoundly warped, as if a strong magnifying glass has been placed where the map represents visual information from the center of gaze. But Inouye's discoveries didn't end there. He reported tantalizing evidence that V1 is not alone; additional visual maps are buried in the human brain.

MAPS OF YOU

In charting the intimate link between the location of damage in the brain and the location of blindness in perception, Inouye revealed the first known visual brain map. Still, the very notion of having a visual map in your brain—not to mention many of them—might strike you as nonsensical. This may be because of your everyday experience with geographic maps. From tourism brochures to mall directories to subway signs, these visual displays are all around us. Like all maps, they are spatial representations of something else. These particular examples are made from *stuff*—actual physical materials such as pigments on a sheet of paper formed from wood pulp.

Of course, many of the maps we see today are displayed on screens, which illustrates how little a map's materials matter. When you pull up a map of your travel route on a computer screen, your route is represented in wavelengths of light emitted from the screen. If you print out the map to take it with you on the road, you re-create the same map with ink on paper, but the representation is the same. That is the beauty of representation: it lets us communicate information about entities and phenomena without having to reproduce them. I don't have to rebuild the pyramids to show you their arrangement at Giza. All I need is a pen and a scrap of paper, a fingertip and a foggy window, or a stick and a sandy beach. In short, it doesn't matter what a map is made of. A map can be fashioned from virtually any material.

Brain maps are not drawn on paper nor lit on a screen; they are made of cells. The brain contains a variety of cells, half of which are neurons. Neurons are interconnected by beautiful branching limbs that carry electrical and chemical signals from one neuron to the next. A neuron can fire electrical impulses one after another, and the rate at which it issues these impulses depends on the information the neuron is representing.

When I think about these rates, called firing rates, I can't help but think of a classroom of young children vying for their teacher's attention: *Me me me me, pick me! Ooh ooh ooh, me!* The more frequently they call out, the more urgency they lend to their message, be it the answer to the teacher's question or a request to use the restroom. The calls from different children in the classroom, like the impulses sent by different neurons in the brain, may have entirely different meanings. But in each case, the frequency or rate of the signal reflects the urgency or importance of the message. When a neuron increases its firing rate, issuing a rapid volley of impulses, this means it has important information to convey at that moment.

Imagine that we could open up someone's skull and unfold the ridges at the back of the brain, so that V1 lies flat. That flat surface of brain stuff would be made up of neurons, just as paper is made up of wood pulp. The sheet of neurons is like the sheet of paper upon which a conventional map is printed. But instead of using inks that vary in hue, brain maps represent information through the firing rates of the neurons they are made up of: which ones are firing rapidly and which ones are barely firing at all. Technically, a neuron's firing rate is the number of electrical signals it sends in a certain amount of time. In essence, you could say that in brain maps, electricity and time play the role that ink does in a conventional map.

Cells, electricity, and time. Those are the raw materials your brain needs to build a map.

The notion of such a map might take a moment to sink in. It is a dramatic departure from more familiar ones. But the map in your V1 is fundamentally no different from one you might keep in your glove compartment. Just as you can translate a map from your computer screen to a piece of paper by printing it, you can translate that same map from the paper to the neurons in V1 simply by gazing at it. One is as good as the next, and all are real maps.

Brain maps are also different from traditional geographic maps because they are dynamic. Once a geographic map is sketched on papyrus or etched on a placard, it is fixed and unchanging. This isn't a big setback because landmarks are immobile, and geographic change tends to happen slowly. When such changes do happen, printed maps become outdated. They can't be automatically updated to reflect a changing world. And so all we can do is discard the old maps and make new ones from scratch.

But some maps *can* be updated. Just think of the map that may be displayed beside your dashboard in your car, or perhaps on the screen of your cell phone. Computerized maps can be updated to represent a newly opened shopping mall or a highway exit that is closed for repair. These maps also incorporate GPS technology, which pinpoints your current location on the planet. From this, your dynamic computerized map is updated as you move about in the world. As you drive north, the map on your display shifts north as well, so that it is always showing you the landmarks that are directly around you. Such a map would be wholly disorienting without the context of your journey and a familiar, salient frame of reference: you. Although your GPS-linked display is constantly changing, or being updated, as you drive, it is still very much a map. And since it has a consistent frame of reference (your current position on the planet), you have no trouble making sense of this dynamic map.

The map in V1 is also dynamic. As you move your body from place to place, as you move your eyes to glance around a scene, and as the objects around you move, the information represented in this map will be updated. But, like your navigation display, the changing information represented in V1 isn't disorienting because it too is anchored to a familiar and salient reference: where you are directing your gaze.

Unintuitive as it may be, maps can be made out of brain cells, and they can be updated and change. But there remains a third challenge to the concept of maps in a brain. Even a map drawn on a foggy window or carved into the sand can be seen. But V1 doesn't turn blue when you stare at the ocean or darken in squares when you look at a checkerboard. Doesn't a map need to have features you can see?

The answer to that question is no. To see how and why, let's indulge in a brief thought experiment inspired by the history of spy warfare. Although we tend to think of ciphers and spy communications as a modern invention, invisible inks have been used for hundreds of years to send secret messages. During the American Revolution, George Washington and his spies used an invisible ink made from a special recipe; this ink could be revealed only when the document was treated with a counterstain. Intelligence, plans, and yes, probably maps penned in this invisible ink were transported without detection and changed the course of the war.

Imagine that one of Washington's spies had used this ink to draw a map of occupied New York City, marked with the gathering places of British troops. Was this invisible map actually a map? Absolutely. And George Washington could have proved it by brushing the paper with the counterstain to make the ink visible. The information contained in the map did not change when the counterstain was applied. The map represented New York City both before and after it became visible to the naked human eye.

The tale of Washington's invisible ink raises an intriguing question: could one apply a counterstain to the V1 map and make it visible? In 1988, a group of vision scientists did exactly that, revealing the V1 map of macaque monkeys. Like humans and other primates, macaque monkeys rely heavily on their sense of sight and have V1 maps similar to our own.

In the experiment, each monkey stared at a flashing pattern on a computer monitor while a compound resembling sugar, but containing a radioactive tag, was injected into its bloodstream. The most active neurons in the monkey's V1 map took up the radioactive compound—this is because neurons that are firing at a rapid rate need more energy. Each monkey was then anesthetized until its heart stopped beating, at which point the scientists added preservatives to the brain tissue, removed the brain from its skull, and removed the

FIGURE 3. A demonstration of the correspondence between the right half of what a creature is viewing (pattern at left) and how that information is represented with activity in the left half of the V1 visual brain map (photograph of brain slice at right). ***From*** **The Journal of Neuroscience,** ***vol. 8, no. 5. Copyright © 1988 by the Society for Neuroscience.***

visual cortex from the rest of the brain. They unfolded V1 so that it lay flat on a slide, froze it, and sliced it with a frozen blade. Then they sandwiched X-ray film against the frozen slices for two weeks to three months before they finally developed the film. It revealed remarkable images of the patterns the monkeys were staring at some weeks or months earlier, in the moments before they died. You can see an example in Figure 3: on the left, the pattern the monkey was staring at and, on the right, the pattern of activity in the V1 map made visible in a slice of the animal's brain.

Just as General Washington added his counterstain to make visible the letters and maps he received, the scientists were able to make the V1 map visible by treating, unfolding, freezing, and filming the brain. In other words, yes, it is possible to open up the brain and see the map in V1, but only with a great deal of effort. New technologies have provided us with easier ways of visualizing brain maps. Ultimately, any method that can translate neurons' firing rates into visible wavelengths of light will do the trick.

Even a cursory inspection of the images above reveals an obvious

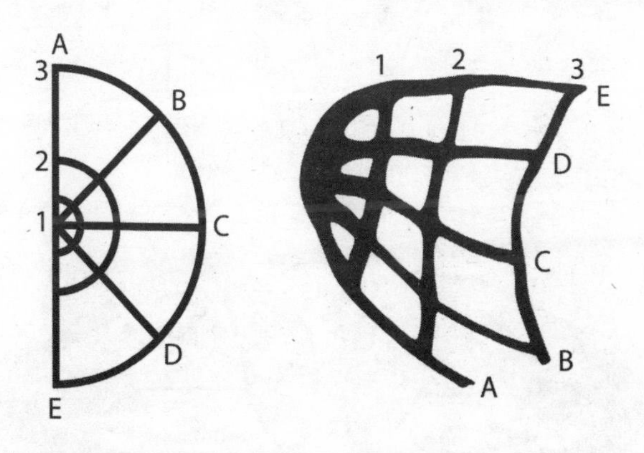

FIGURE 4. Labels on the viewed visual pattern (left) and on the corresponding activity within the monkey's V1 visual map (right) reveal how the representation in the V1 map is inverted and warped. *Paul Kim, adapted from* The Journal of Neuroscience, *vol. 8, no. 5. Copyright © 1988 by the Society for Neuroscience.*

problem: the pattern on the brain slice looks quite different from the pattern on the display. The difference is not an error. The activity in the monkey's V1 map was not a faithful representation of what the monkey was staring at before he died. The V1 map is profoundly warped. Labels on the panels in Figure 4 reveal how. The vertical straight line on the left side of the original pattern has been stretched into a wide C-shape in the brain, while the perfect semicircle on the right side of the pattern has been flattened and even slightly inverted. The pattern is upside-down, so that the upper portion of the display is represented in the lower part of the V1 map. And there's more: something is wrong

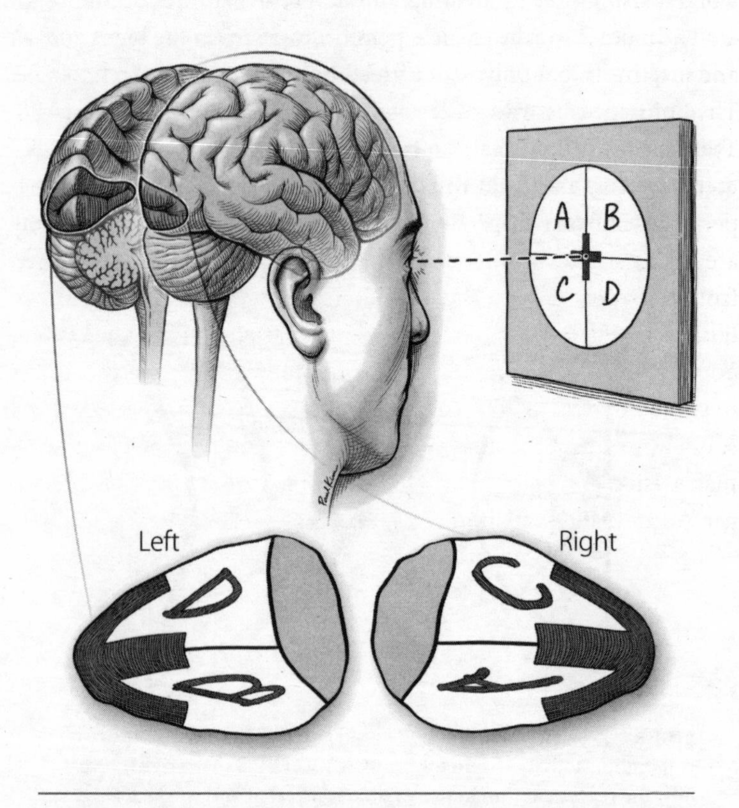

FIGURE 5. How visual information is represented on the left and right sides of the human V1 visual map. ***Paul Kim***

with the brain's representation of the concentric circles, so that the smallest semicircle takes up far too much space. These anomalies are the very distortions that Inouye discovered more than a century ago.

Thanks to the work of Inouye and a host of scientists before and after him, the once elusive seat of visual perception in the brain is known. The region is tucked away within folds at the back of your brain. It contains a map made of neurons that represent visual information by means of electricity and time. The illustration in Figure 5 shows where V1 hides and how its visual representation is laid out. This is the map that was pierced by bullets in Inouye's patients, leaving them with holes in their vision even though both of their eyes worked just fine.

The mere existence of such a map in your head may seem strange and improbable. Yet maps like V1 are not the exception but the rule. The brains of creatures great and small are chock full of such maps. The coming chapters will explore their remarkable variety and how their idiosyncrasies and distortions shape what you think and experience. But first there is a fundamental question to answer: why are brains such consummate mapmakers? The answer can be culled from electronics design and evolution, and it touches on topics from hungry brains to the navigational prowess of the humble desert ant. You will come to see that what is truly improbable is your capacity to see your world at all. Visual brain maps like the one in V1 are the solution to a problem that you never knew you had. They uniquely make vision and your other senses possible in a world of bitter hunger, scarcity, and predation.

2

The Tyranny of Numbers: Why Brain Maps Exist

THE ENGINEERS OF BELL LABS were not interested in brains. They were interested in designing devices that met people's needs. But by the late 1950s, Bell Labs and the nascent electronics industry faced the same problem that has plagued brain evolution for millions of years and made brain maps a biological imperative.

The vice president of device development at Bell Labs at the time gave the problem a name: the tyranny of numbers. An electrical device operates because of its electrical components, or the elements within it that support its essential functions. Consumers wanted more powerful and more versatile gadgets, so that a single device could carry out more than one function. Therefore engineers tried to design new devices packed with a skyrocketing number of components. It was hard enough to figure out how to pack millions of elements into a reasonably sized gadget. But these additions created an even bigger problem: for each new element added, the engineer had to include many new wires to connect it to the other elements in the device. This difficulty was referred to as the tyranny of numbers. Adding power and functionality to a device called for more elements, but more elements meant more wires, which raised manufacturing costs and took up lots of space. The result? Hulking machines with astronomical price tags.

The solution to the tyranny of numbers would come from elsewhere. Jack Kilby at Texas Instruments devised a solution: an integrated circuit, which allowed engineers to incorporate many elements within a single piece of germanium metal, dramatically reducing the number of wires needed. Robert Noyce at Fairchild Semiconductor in Mountain View, California, invented a silicon version of the integrated circuit that would prove to be the namesake and cornerstone for Silicon Valley as we know it today. The integrated circuit made it possible to pack far more elements into a single device. This innovation would drive the modern age of electronics and give rise to the powerful multifunctional portable devices that dominate modern life.

But the tyranny of numbers never entirely went away. Fast-forward to today, and consider the cell phone that probably accompanies you everywhere you go. Many cell phones are multifunctional, mini–computing devices: a phone, a camera, and a player of music, movies, and video games. The modern cell-phone chip contains billions of transistors and integrates numerous elements that enable it to fulfill these many roles. Yet it must also be light and small enough to fit into your pocket or purse. These conflicting pulls—more features in smaller packages—will create headaches (and job security) for engineers well into the future.

Much of the process of designing a modern cell phone comes down to a handful of questions that are equally relevant to brains. What must the capabilities of the final product be in terms of the functions it can perform, its capacity to process large amounts of information, and the speed with which it can do so? What does each part of the device do, and how do these parts need to communicate with one another? How much will it cost to construct this configuration? How small and lightweight does the final product need to be?

Unlike devices meticulously designed by engineers, brain structures and the capacities they support are shaped by natural selection over generations. No one is consciously weighing design options for the brain; genetic mutations, reproduction, and death over generations together optimize a creature's basic brain structure through blind trial and error. Still, it is easier to consider the tradeoffs inherent in brain evolution if you view it as an engineering challenge. *What does it take to engineer a brain?* The answer depends on what the brain must be able to do.

When people talk about what a brain can do—its capacity to process information and support intelligent behavior—they tend to talk in terms of *better* or *worse*. By and large, they assume that every creature and each person can be assigned to a particular rung on an intelligence ladder, with *dumb* down in the dirt and *brilliant* up in the clouds. But a closer and less biased assessment of animals' capabilities reveals how flawed that assumption truly is.

Consider the desert ant, which regularly navigates the harsh Sahara in search of tiny edible morsels. Relative to their size, these ants trek distances that would be on the order of miles for us, and then find their way back in the all-but-featureless desert to the pinpoint location of their colony's nest. Or consider killer whales, which live in groups called pods and communicate using calls with a dialect that is unique to their specific pod. When the pods disperse and the member whales roam several miles apart, they must "tune in" to other whale calls in their dialect and ignore those from other pods. This allows them to track the locations of their pod mates and eventually reunite. And then there's Clark's nutcracker, a bird that gathers pine seeds each autumn and buries them in small holes for safekeeping. The bird remembers and visits well over three thousand of these hiding places during the winter and spring to retrieve food for itself and its offspring.

The brains of these animals range in weight from about 0.001 grams (ants) to about 3,650 grams (killer whales), but they are all capable of impressive cognitive feats that would either challenge or prove impossible for us and our own roughly 1,500 grams of brain stuff. A surprising variety of living things can be called smart, but they are smart in particular ways that help them overcome specific challenges. The sheer variety of abilities supported by brains on this earth makes it impossible to compare them based on any single metric.

Across species and lineages, natural selection has helped produce this variety. Genetic changes can create changes in brain structure that, in turn, support new abilities or behaviors. There are many ways to tinker with a brain. Brains can be larger or smaller, of course. But brains can also differ in neuron size and density, or in how many neurons can be packed together in a given unit of space. For example, elephants and whales have enormous brains containing large, widely spaced neurons, whereas great apes have smaller brains that are jam-packed with smaller neurons. Brain size and neuron density together determine how many neurons a brain contains. The cow and the chimpanzee have brains that are about the same size, but since the chimp's brain is neuron-dense, it is believed to contain far more neurons overall. That matters because the number of neurons in an animal's brain is a crucial factor for its survival. Having either too many or too few to meet its needs is a veritable death sentence.

Imbuing a brain with extra neurons imparts obvious benefits. Since neurons are the units that process information, increasing their number can increase the brain's processing capacity. A sizable chunk of most animal brains is devoted to processing the information that comes in from its senses—in our case, information from vision, hearing, touch, and the like. Having a pair of functional eyes is not enough to endow you with your sense of sight; you need brain areas like V1 to represent and process the torrent of information that those eyes collect. With additional neurons, a brain can support better sensory perception, boosting the creature's ability to detect food or predators. Alternately, those extra neurons could support the creature's ability to make more complex or accurate movements, enabling it to better succeed at catching food or evading predators. It could also support other capacities, such as navigation, memory, self-control, planning, or reasoning, any of which could be a boon for a creature's survival.

But additional neurons also come with some very high costs. First and foremost, neurons are profligate consumers of energy. For example, the brain ranks third among human organs in terms of energy consumption per unit weight, after our ever-beating hearts and assiduous kidneys. But since our brains are larger and heavier than either of those organs, they consume the most energy overall. This one chunk of tissue makes up a mere 2 percent of the adult human body weight but guzzles about 22 percent of the energy you take in. In other species, these percentages will differ; but across the animal kingdom, neurons demand a great deal of energy. Neurons' energy price tag is this high because they must literally pump certain molecules into themselves and pump certain molecules out of themselves to render them capable of sending even a single impulse. This pumping is happening constantly in neurons throughout your brain. The more neurons a creature has, the more miniature pumps it will need, working night and day and requiring plenty of fuel.

Since neurons are so expensive, brains with more neurons demand more calories. Creatures can meet this demand by eating more food *or* eating better food that is higher in calories. Scoring high-calorie foods may not sound like a problem in the context of our modern industrialized world. In many wealthy nations, grocery store shelves are packed with inexpensive processed foods. But that is only a recent development—a veritable eyeblink in the time line of human evolution. Throughout most of the time that our ancestors have roamed the earth, they faced the stark reality that wild animals do today. There is only so much edible, nutritious food to go around. Every calorie is hard-won. And if your body and brain require more energy than you are able to scavenge, hunt, or steal, well then, your time is up. So long and goodbye.

Neurons are also costly in terms of space. Every neuron must communicate with other neurons, and they do this through protuberances called dendrites and axons. These tentacle-like extensions serve as wires that convey information in the form of electrical pulses from one cell to another. These wires have highly specialized roles. The dendrites of a neuron receive incoming signals, while the longer axons convey each neuron's own signals to the dendrites of other cells. Effectively, dendrites are the ears of a neuron and the axon is its voice.

Adding neurons to a brain can increase its capacity to think, perceive, and act, but only if those neurons are wired up to communicate with one another. The wirelike axon in particular supports this crucial function. Axons can carry a message from one neuron to its next-door neighbor or transmit impulses all the way across the brain. Sensory receptors like the touch receptors embedded in your skin are also a kind of neuron and have axons of their own, which they use to transmit signals to your brain. There are even axons that span the length of your body; they carry touch information from toe to brain, allowing you to experience the pleasures of a foot massage or the pain of a stubbed toe.

Axons are essential for the brain to serve its many functions. But they eat up a lot of space. In fact, when evolution has added neurons to brains, the new connections required actually take up more space than the new neurons themselves. Longer wires use up more space than shorter ones, and maintaining them requires more effort. The engineers at Bell Lab faced a similar challenge decades ago: adding new elements to a device can imbue it with more capabilities, but it also entails adding lots of bulky and expensive wires. What engineers called the tyranny of numbers we might call the tyranny of neurons. And if it sounds like a minor problem, think again.

Neurons come in a splendid array of types, each with its own properties, but a single neuron needs to directly connect with hundreds of other neurons in the brain in order to carry out its functions. Without a clever solution to this design challenge, costly wires would gobble up space and energy in brains, leaving creatures with heads too large to lift and caloric needs that would drive them to starvation. Your brain contains about 86 billion neurons. If each were randomly connected to all of the other neurons in your brain, that organ would have to be more than 20 kilometers in width to house all of the connections—despite the fact that each axon is thinner than a human hair. Luckily, neurons do not need to connect to all other neurons; each relies on connections with a small fraction of the 86 billion other neurons in the brain. Even so, any brain comprised of randomly connected neurons would be impractically large.

Electronics design offers a model of how to make this work. Designers place those components that need to work together and share information as close as possible to one another in a device. This saves space by reducing the length (and therefore volume) of wires needed to connect the components. The same idea applies to brains. If two neurons rely on each other to do their job, they'll need to be interconnected. It saves energy and keeps wires short if those neurons are situated next to one another in the brain.

A single neuron in the brain needs to talk with other neurons—but which ones? Recall the fragmented nature of your sense of touch, which begins as a mosaic of touch information collected by individual sensory receptors in your skin. I described a touch receptor buried in the skin of your right kneecap as a reclusive landowner, caring only about its little plot of kneecap territory. That territory is its receptive field. If something presses against this swath of skin, the receptor sounds the alarm by firing a rapid volley of impulses. *Something is happening on my land!* These signals are sent on to the brain to neurons that also have receptive fields. In the brain there are neurons whose activity represents pressure on your right kneecap, even though there are no actual kneecaps in the brain. So who does a right-kneecap neuron in the brain need to talk to most? Other neurons in the brain that represent touch on the right kneecap. And who else? Neurons that represent touch on the right upper shin or on the right lower thigh. In general, *neurons need to talk most with other cells that represent the same things or other things nearby.*

There is a reason why neurons must talk most with their like-minded kin, and this reason has a name: local processing. Local processing is essentially comparing the situation at one point in space to the situation in neighboring points in space. These could be points in space on your body, in the case of touch, or in your visual field, for seeing. In essence, local processing involves seeking answers to the following questions: *Is something happening here but not there? There but not here? Everywhere? Or nowhere?* These specific comparisons sound tedious, but they are terribly important. They provide information about precisely *where* something is happening, such as a painful sensation on your kneecap. These comparisons also provide valuable context for processing other sensory inputs. For instance, receiving information about a painful event taking place on one part of your kneecap could help other neurons, representing other patches of skin, to detect painful events as well. This type of communication is especially important between neurons that have neighboring receptive fields. In neural representation, as in your own everyday life, events that happen nearby tend to be the most relevant to your own situation. If the house next door to you is on fire, you had better know about it. A house burning on the other side of the planet? Not so much.

Often, comparing your situation to that of your neighbor's also yields valuable clues as to *what your situation is.* For example, imagine that the electricity in your house or apartment goes out. If yours is the only property affected, you might call an electrician for repairs or check to see if you paid your latest bill. But if your neighbors have lost their power too, the cause of your problem could be miles away, and there is little for you to do besides wait for municipal services or the energy company to fix it. A similar situation faces a humble neuron representing pressure against your right kneecap. There is pressure on your kneecap whenever you wear a particularly tight pair of pants or stretch out on your stomach for a snooze, but in both cases this pressure affects many parts of your body. There is no need to become aware of the sensation on your kneecap in particular. Compare that to the touch of someone's hand upon your knee or the pressure of a ball colliding with it. These pressures happen only on your knee and signal that something unique and knee-relevant is happening there. *Stop everything and notice that knee!* If our little kneecap neuron is going to help us tell the difference between these events, it will need to talk to neurons that represent pressure on the skin above the kneecap, below it, and on either side of it.

Now imagine two neurons that represent touch information in the brain: one with a receptive field on the right kneecap and another with a receptive field on the right upper shin. Each would do well to know what is happening on the other's turf. They'll need to communicate and compare information, and that means they'll need to be wired up to talk. If you put the kneecap neuron next to the shin neuron in the brain, they can gab to their hearts' content using short little wires. Of course, kneecaps and shins are only two zones within a grander skinscape. A kneecap neuron must also talk to lower-thigh neurons, whereas shin neurons must talk to calf neurons and ankle neurons. In order to keep wires short, each of these neurons must also neighbor one another. This principle extends to the arms, neck, and face, in one direction, and to the toes in the other. Neighboring neurons in your brain represent neighboring plots of land on your skin. The result? A beautiful, honest-to-goodness map of the surfaces of your body *built into your brain.*

There are many such maps of your body in your brain. One of the best known is called the primary somatosensory cortex, or S1. Just as V1 is the first region of the outer surface of your brain, or cerebral cortex, to receive incoming information from light receptors in the eyes, S1 is the first region of cerebral cortex to gather information coming from touch receptors. This region is located at the very top of your brain. And it is laid out according to the surfaces of your skin, complete with your tongue and lips, your nose, two eyes, ten fingers, one tummy, two kneecaps, and ten toes. This brain map of your body allows you to devote as many neurons as possible to detecting touch while still keeping the wires between them short and your hat size and appetite small.

This elegant solution also explains the visual map that Inouye found in the brains of his wounded soldiers. Cells that represent light striking neighboring points on the retina should also be cuddled up close together in the brain, allowing them to share information using the shortest possible wires. This sort of information sharing is what allows your brain to swiftly and accurately detect important boundaries in a visual scene, such as where lines or surfaces end. These boundaries provide essential clues, revealing *where* the objects around you begin and end and, crucially, *what* those objects are.

This talk-to-your-neighbors principle also explains why brain maps exist for other capacities, such as hearing and moving, and are found in creatures throughout the animal kingdom. Maps allow evolution to "soup up" the brain with extra neurons while keeping its overall size and energy needs under control. Brain maps also benefit creatures that have fewer neurons by keeping their brains as small and energy efficient as possible. Thanks to maps, these creatures can make the most of the neurons they have, leaving them nimbler and less likely to starve. In short, nature has hit upon a design scheme that is good for brains complex and simple, big and little alike. Brain maps overcome a major obstacle to survival. In this world of finite resources and stiff competition, they make life possible for desert ants, killer whales, and you.

HOW TO DARN A MAP

There are plenty of reasons to be grateful for your brain maps. The alternatives—starvation, immobility, and extinction—are none too appealing. You can thank brain maps for the speed and clarity of your senses, not to mention the fact that you have the headspace to harbor *five* senses instead of one or two. But brain maps offer other benefits as well. They are perfectly laid out to detect and correct the inevitable errors and omissions in the information your brain receives from your eyes, ears, and skin.

A wonderful demonstration of how brain maps correct perception comes to us from a humble priest and self-taught scientist of the seventeenth century. Edme Mariotte was abbot and prior of St. Martin de Beaumont-sur-Vingeanne, a small village in the countryside near Dijon, France. Over the course of Mariotte's life, he would study topics ranging from plant physiology to physics, astronomy to anatomy. He would organize one of the first international scientific collaborations and earn a complimentary mention in Isaac Newton's masterwork, *Principia.* But his most famous contribution to science was his discovery that we are all a little bit blind.

Mariotte participated in numerous dissections, poring over corpses of animals native to the French countryside, exotic animals when available, and even humans. He was particularly interested in the anatomy of the eye. Along the back of each eyeball—animal and human alike—he found a sunken oval that stood out from the rest of the retina. He was not the first to notice this oval, which was known to be the place where the optic nerve that carries information from the retina to the brain exits the eye. The oval was called the optic disc, and its role in vision was unknown. Scientists of the day gen-

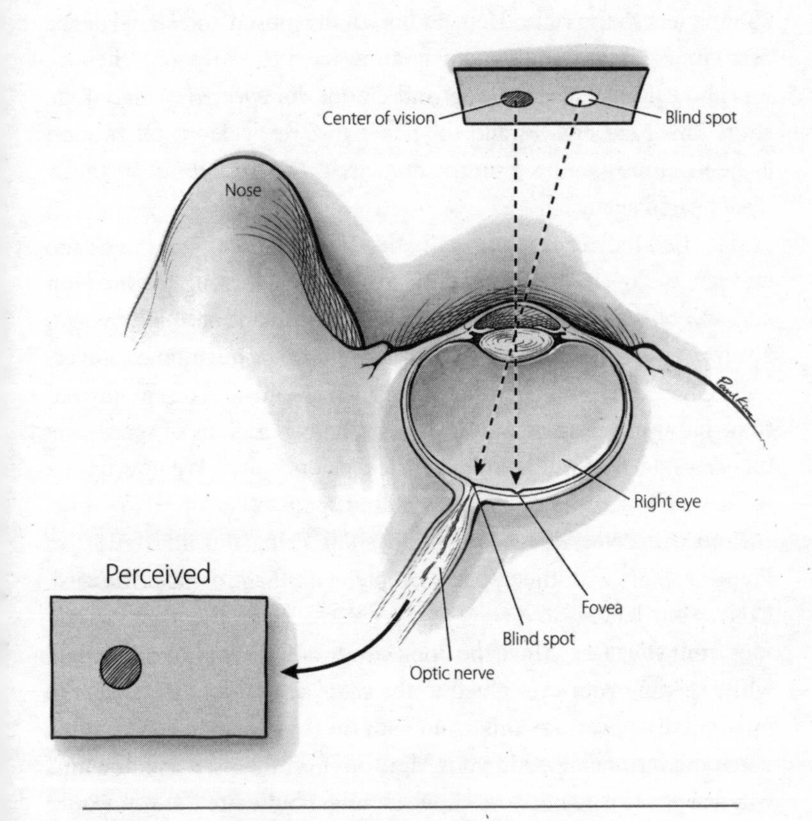

FIGURE 6. An illustration of Mariotte's experiment. ***Paul Kim***

erally believed that the optic disc was the place of clearest vision in the eye. However, in Mariotte's dissections, he noticed that the optic disc never lies in the center of the back of the eye (corresponding to central vision, where our vision is best). Rather, in humans, it was slightly higher and closer to the nose. Mariotte was curious about this discrepancy. And so, with nothing but a few paper cutouts, he set out to test the nature of vision at the optic disc.

Mariotte's test ranks among the simplest scientific experiments ever undertaken. The diagram in Figure 6 illustrates how it worked. He attached one small white paper circle at eye level on a dark background, and a second circle, about 4 inches across, a little lower and about 2 feet to the right. He held his left eye closed and stared at the first circle, while stepping slowly away from the display. When he was about 10 feet away, the second cutout *disappeared*. Amazed, he shifted his gaze slightly, and the paper disc reappeared. Yet as soon as he fixed his gaze back on the first circle, the second one instantly disappeared again.

He tried the experiment again, this time with his right eye closed and left eye open, and at different distances, adjusting the location and size of the second circle accordingly. The spot of blindness was always there. He tried out the experiment on an acquaintance, Reverend Billy, and eventually on other French scientists. They all proved to be blind in the same way, always in the two regions of space, one on each side, that would fall on each eye's optic disc. We now have a name for these zones of blindness: blind spots.

You can detect your own blind spots using the illustration in Figure 7. Start with the upper example, with the cross and the bird. Cover your left eye and stare at the cross, placing the page about a foot from your face. Move the book closer or farther away as needed, while keeping your eyes glued to the cross; at the right distance, the bird will disappear. Try this again with the lower example. This time, when the bird falls within your blind spot on the right eye, the bird will disappear while the cage appears intact but empty. If you would like to try this using your left eye, just turn the book upside-down, cover your right eye, and repeat.

FIGURE 7. Demonstration of the blind spot. Use these illustrations and the instructions in the text to reveal your own blind spots. ***Paul Kim***

We now know that the optic nerve is the bundle of axons that carry outgoing messages from the retina to the brain. These axons take up space where they exit the eye, leaving no room for light receptors to collect information about incoming photons of light. And so every human eye has an optic disc, and every optic disc creates an oval of actual blindness.

Simple as Mariotte's experiment was, his blind spot became the talk of Europe and sparked new scientific debates about vision and the eye. It also raised intriguing questions: Why aren't we aware of these blind spots? Why don't we notice them in everyday life? Mariotte ventured a couple of solid guesses. We usually view the world with two eyes simultaneously, so that the region of space that we don't see with one eye will be seen with the other. Moreover, we naturally shift our gaze quite frequently, which means that no single part of a visual scene would stay hidden in a blind spot for long.

Both of these points are valid, but neither explains why Mariotte's blind spot was not apparent to him when he had one eye closed and the other unmoving. They do not explain why Mariotte's second circle disappeared when it fell within his blind spot, to be replaced by

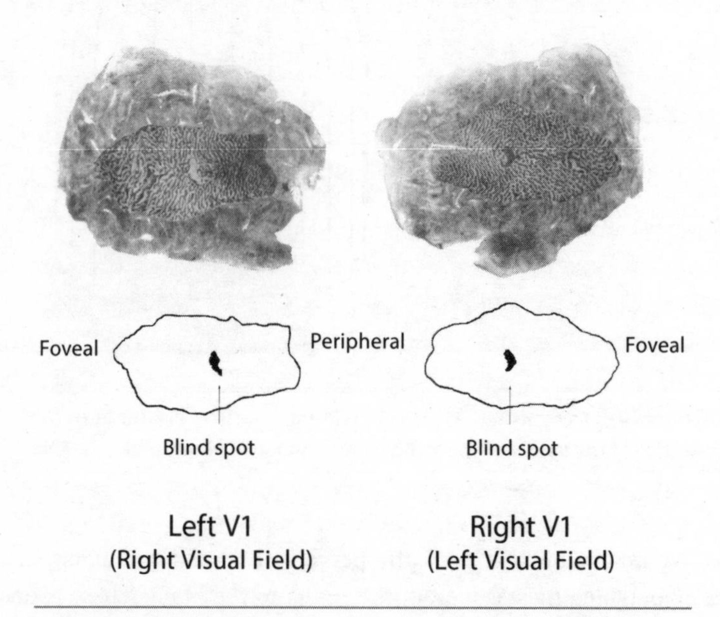

FIGURE 8. Photographs (above) and outlines (below) of a human V1 visual brain map. The photos have been stained to make the map's features visible. The regions of V1 that correspond to the person's two blind spots are clearly visible in the photographs as light and dark islands within the left and right halves of V1, respectively. The distinctive banded pattern visible elsewhere in V1 reflects input from the two eyes. (See Chapter 11 for more detail.) ***From* The Journal of Neuroscience, *vol. 27, no. 39. Copyright © 2007 by the Society for Neuroscience.***

the dark background, just as the grid of the bird's cage remained intact while the bird within it disappeared from your view. When the bird disappears, it does not seem to vanish into darkness. It appears to be replaced by *something*—the white background of the page or the lines of the cage. Why should that be so? This question would go unanswered for centuries, even as the world discovered electricity, receptive fields, and brain maps. Ultimately, the answer would be found within the living V1 map.

Recall that your V1 map is arranged according to the layout of the retinas in your two eyes. Although each eye independently collects information about incoming light, the information from the two eyes is combined into one big map when it reaches V1. The V1 visual map is split between your right and left hemispheres, or halves of your brain; the half of V1 in your left hemisphere represents information from the right half of your visual field and vice versa. In Figure 8 you can see pictures of actual slices of a human brain, showing both the left and right halves of the V1 map. After the person had died, this tissue was removed from the brain, flattened, and stained with a compound that made some of the features of the map, including its boundaries, visible. These stained brain slices also show the regions of the map that represent the blind spot of each eye. For clarity, labeled outlines have been placed below the slices. The large patch indicated with the arrow in the left hemisphere represents the part of the visual field that falls in the blind spot of the right eye. The comparable patch shown with an arrow in the right hemisphere represents the part of the visual field that falls within the blind spot of the left eye.

Most of the time, you have both eyes open and all is well in these regions. The part of the map that corresponds to the blind spot of your right eye still gets visual information about that part of visual space from your left eye, and vice versa for the blind spot of your left eye. But when you close one eye, as Mariotte did for his experiment, one of these blind-spot regions in the V1 map no longer receives messages from any eye. This region of the map is still living, active tissue in the brain, but it has been cut off from its usual source of visual information. Yet rather than perceiving a patch of blindness, as Inouye's patients did with their scotomas, you perceive *something*, completing or filling in the spot so that it resembles surrounding space.

Exactly *how* the brain manages this perceptual filling-in remained a mystery until recently, when experiments with monkeys offered new answers. Using tiny electrodes, neuroscientists measured the activity of neurons that represent one of the blind-spot regions in the monkeys' V1 maps. Like Mariotte and other humans, monkeys experience their blind spots as filled in with surrounding colors or patterns when the opposite eye is covered. The scientists discovered that a special set of neurons in the patch of monkey V1 corresponding to one eye's blind spot becomes active when the animal's opposite eye is closed, cutting off visual input to the region. These special neurons have enormous receptive fields; each represents part of the blind spot and also part of the visual space surrounding the blind spot. In essence, these cells picked up signals from the part of the retina surrounding the blind spot and stretched them to fill in the representation within the deprived patch of brain. Just as a hole in a sock can be mended by darning, or stitching the edges of the hole together, the blind spot can be filled in by these very special cells.

This filling-in, or darning, of the blind spot in V1 is a dramatic example of what scientists call perceptual filling-in. Based on the incomplete details the eyes have gathered, the brain uses the information it has to fill in information that is missing. And this is only one of many forms of filling-in that happen in our sensory experience. Many optical illusions work because of our visual system's natural tendency to fill in information that seems to be missing, even when nothing actually is. For example, if you look at two moving stripe patterns that are aligned but separated by a gap, you'll perceive that the stripes extend across the divide to connect with each other, even though they do not. A team of scientists tested this illusion with people to examine what was happening in the region of the V1 map that represented the gap while they were experiencing the illusion. Did activity in this region reflect the information that the eyes were collecting (that is, no movement in the gap) or the perception of moving stripes filling in the gap, which is what the participants experienced?

To observe the V1 maps in living humans, they used a popular type of brain scan that can be performed with an MRI machine. Called functional MRI, it gives scientists a sense of how brain activity changes over time or in response to experimental manipulations. When participants undergo functional MRI scans while they watch certain bright, flashing visual displays, scientists can use the results to see their V1 maps of the visual field. In this way the scientists studied activity in the human V1 map while people looked at the moving stripe patterns and experienced the illusion. Activity in the part of the V1 map that represents the stationary gap ramped up when people experienced the illusion and perceived moving stripes spanning this gap. In short, the illusory stripes completing the movement across the divide could be detected in the V1 map itself. The stripes were extending across the gap both in the test subjects' V1 maps *and* in their conscious visual perception.

Perceptual filling-in happens all the time, although we are virtually never aware of it. Most of the time, it goes unnoticed because the brain fills in information correctly based on the visual data it receives and thus steers us well through the environment. Carefully designed visual illusions, however, give us the chance to experience how our own perception is continuously undergoing revision. But why does the brain go to the trouble of filling in information in the first place? A likely answer is that we benefit from having a visual system that anticipates continuity. For example, the visual system that you enjoy evolved and developed to make sense of the three-dimensional jumble that is our world. Your view of more distant objects is regularly obscured by closer ones. When you walk into a room for the first time, your visual system could, and by all rights should, be overwhelmed by a disorienting rush of new lines, angles, colors, textures, and shadows.

In truth, we don't usually care about these lines and textures and other details—not really. What we care about is the objects beneath them. To help us see these objects, the brain needs to factor in what is probable. For instance, there will likely be no random holes in space, and no uncanny coincidences. In the case of the illusion with the moving stripes, what is the likelihood that the two separate moving stripe patterns just happen to be perfectly aligned and synchronized? Aren't the odds greater that they would be two parts of a single large pattern, with a strip of something else obscuring your view? So we process the stationary gap between the moving patterns as something irrelevant, in the way. We mentally represent the moving lines across the divide, opting to represent the *stuff* that is probably there, rather than the angles, lines, and textures that our two eyes dutifully report.

Consider another example. A narrow cloth runner draped from one end of a tabletop to another is merely that—cloth on top of a continuous table, and not two separate tables with a gap in between. Thankfully, we don't perceptually fill in the cloth runner with tabletop; we are able to perceive *both* the cloth *and* the single table beneath it. Still, functional MRI studies of humans indicate that the V1 map does represent information about objects or portions of objects that are currently obscured behind closer objects. In short, the perceptual filling-in that we experience through many popular visual illusions may be a dramatic example of a subtler and more essential phenomenon: based on what our eyes take in, the brain automatically fills in, or extrapolates, what else is actually before us.

Brain maps like V1 are beautifully suited to support this kind of extrapolation because they are continuous and laid out in a way that makes local comparisons easy and quick. Just as local comparisons within a brain map can highlight key places where something important is happening, they can also help identify regions of visual space that are improbably different from their surroundings. Like the autocorrect function on your word processor, brain maps can rapidly identify and correct inputs that appear to be errors. Given sufficient training (or experience) and the right programming (or wiring), your word processor and your brain maps can fix errors and avoid pitfalls without your ever even noticing that they were there.

It is easy to take perception for granted. After all, it comes naturally and effortlessly. You've never had to take a class in perceiving or found yourself burnt out from doing too much of it. Plenty of the things that we do or think about require actual effort, but perception simply happens. It dynamically unfolds every waking moment of our lives from the day we are born until the day we die. And so it is easy to overlook the miracle that perception truly is. Major challenges in terms of space, energy, mobility, and information processing could easily have kept you from ever recognizing an object in your path or sensing someone's hand upon your knee. Brain maps were the key to overcoming physical and computational impediments alike.

And yet, for all of the challenges that brain maps overcome, their mere existence is not enough. To make the powerful yet practical brains found on earth today, these maps must feature gross distortions. As you will see, these distortions are a hallmark of how your brain represents your world. And that, in turn, makes them a hallmark of how you perceive the world and everything in it.

3

How Brain Maps Determine What We See and Feel

DESCEND INTO THE SUBWAY TUNNELS that lie beneath Boston, Massachusetts, and you will find maps everywhere: framed on station platforms, plastered inside train cars, and printed on flyers. For the city's 1.3 million riders, the subway map serves equally as resource and decor. But it also illustrates a surprising fact that is as true for maps in brains as it is for maps in subways: sometimes distorting a map actually makes it better at representing what people need to know.

The lines of the Boston subway system are laid out like the spokes of a wheel, with its hub at the heart of the city, as you can see in Figure 9. Transferring from one subway line to another can happen at four stations in the hub. I once rode the Red Line from one hub station (Park Street) to another (Downtown Crossing), only to discover that I had traveled just a single city block. Compare that to the distance between the Davis and Alewife stations, two adjacent stops at one end of the Red Line. Although the dots representing Davis and

Alewife on the map are closer together than those representing Park Street and Downtown Crossing, the physical Davis and Alewife stations are separated by more than a mile. In short, the map's representation of the subway system is warped with respect to reality. If 1 centimeter's distance between Park Street and Downtown Crossing on the map represents 10 meters of actual track distance, then 1 centimeter on the map between Davis and Alewife represents a whopping 140 meters of track.

Remarkably, this inaccuracy actually makes the subway map more useful. When you are riding away from the hub, you need to know only the serial order of stations along that track in order to an-

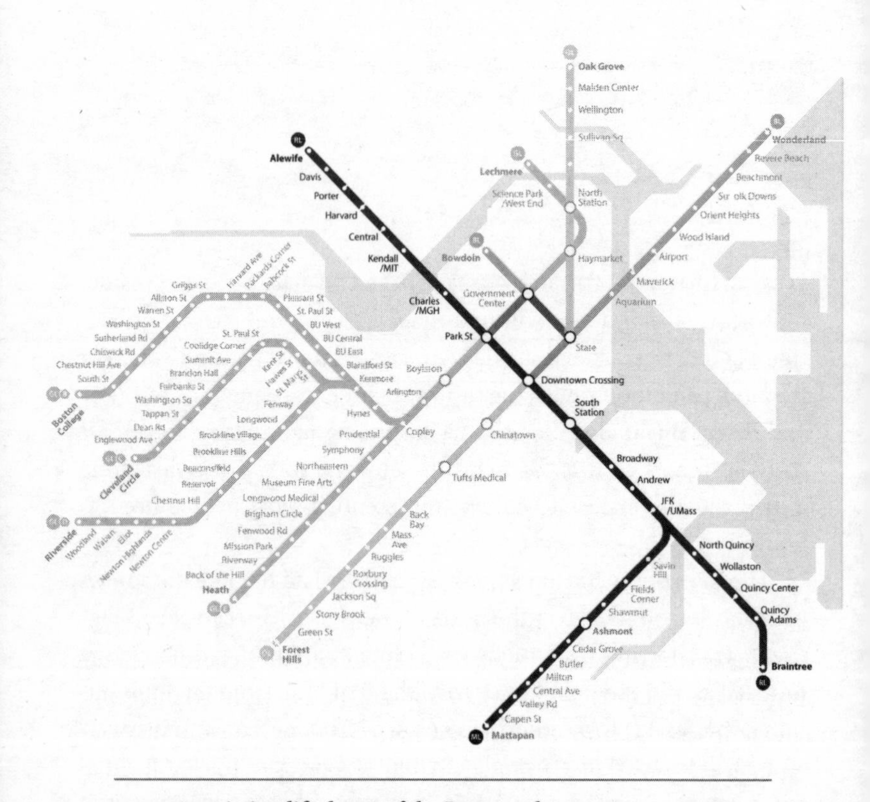

FIGURE 9. A simplified map of the Boston subway system. ***Paul Kim, adapted from Michael Kvrivishvili (CC BY 2.0).***

ticipate and disembark at your destination. But when you're riding toward the hub and preparing to transfer, you also need spatial information. *Can you get to the airport with one transfer, or do you need to make two? At which stations do you make these transfers and, when you reach them, in which direction should you ride on each new line?* By magnifying the representation around transfer points, the subway map gives you spatial clarity where you need it most.

Many brain maps, including your visual map in V1, employ the same trick. The part of the V1 visual map that represents visual information from your foveas, corresponding to where you are directing your gaze, is dramatically magnified compared to the rest of the map. In fact, this magnification lies at the heart of one of Inouye's discoveries: although the bullets for the Mosin-Nagant guns were exactly 7.6 mm in diameter, they created patches of blindness that varied widely in size. Blind patches at his patients' center of gaze were quite small, while those in the periphery were far larger. These size differences show that the representation of foveal information is grossly expanded in the V1 map, just as the representation of the hub is magnified in the subway map.

Why is foveal information in particular magnified in the V1 map? In essence, because it pays to invest in one's strengths. Light receptors are packed more tightly together at the fovea than anywhere else in the retina. Because we have more receptors at the fovea, we collect more information about the part of the visual scene we are looking at directly. This difference creates an inequality within the retina, making the fovea a little better than other zones. But that inequality is only the beginning.

Imagine two of the light receptors in your retina, one located in the fovea ("Florence") and the other off on the periphery ("Perry"). Let's say that these receptors each detect a photon, or packet of light, at the same instant and send the same signals announcing their discovery. In less than one-twentieth of a second, these signals will travel through other cells in the retina, make a brief stopover at a relay station nestled deep in the brain, and reach V1 at the back of your head. Along the way, the signal from Perry is combined with signals from his neighboring receptors, while the signal from Florence is faithfully preserved. Although the two signals were identical when they began their journey, by the time they reach V1 they are nothing alike. There, information from Florence gets a hundred times more space in the V1 map than information from Perry. The fact that input from Florence and her neighbors at the fovea claims grander territories in the V1 map has real consequences for what you can perceive. Larger territories mean more V1 neurons devoted to representing the fine-grained detail of patterned light detected at the fovea.

The story of Florence and Perry raises obvious questions: Wouldn't we see better if we preserved signals from Perry just as faithfully as we preserved signals from Florence? Why magnify just one part of the map when you might magnify all of it? The same questions could be asked about the subway map. Why not give the map a consistent scale and then just print the whole thing bigger, so that it is accurate *and* clear? The answer is that if the Boston subway map represented the entire system at the same scale as the hub, the map would have to take up roughly a hundred times the area overall. Displaying such a map at an underground platform, much less printing it on flyers, would be all but impossible. The subway map is a compromise: it provides detail where it is needed while keeping the overall map reasonably sized.

The same compromise applies to the visual cortex. In a perfect world, V1 could equitably represent foveal and peripheral inputs alike. But you could never afford to have such a map; your V1 would have to be thirteen times larger to accommodate it. Worse still, that extra information would have to be processed in other brain areas, meaning that they too would have to be larger. If the brain were organized in this way, areas dedicated to vision alone would be too large to fit inside the human skull.

Forced to play favorites and make sacrifices, your brain went all in on the fovea, at the expense of the periphery. This sacrifice was possible and even prudent because your eyes are remarkably agile. Humans make about five quick eye movements per second pretty much every waking moment of their lives. These eye movements are so frequent, so familiar, and so beautifully stitched together by the brain that we are typically not even aware of them, although we can consciously detect them when we pay attention. If you try to read this sentence without moving your eyes, you will see the impact of this specialization in action.

We use the fovea as we would a single telescope tasked with collecting information about the entire night sky. We point it here and there, to one place after another, collecting details about areas of interest and then compiling these snapshots into a fuller portrait of the heavens. Making one high-resolution telescope and swiveling it to take consecutive snapshots of the sky gives us a happy compromise between seeing clearly and seeing a lot.

Information from the visual periphery is also valuable to us, but we use it for a different purpose and it has been specialized accordingly. Peripheral vision is not as sharp as foveal vision, but it is good at detecting movement, and it operates well in dark or dim settings. It gives us the coverage we need to detect the unexpected. When Perry and his neighbors pick up a surprising movement in the periphery and send their report on to the brain, your eyes will move swiftly to lock onto the source of the motion. Then Florence and her cousins take over, sending details to help you determine the cause of the motion and whether it is a threat.

Magnification is a mainstay of brain maps. It reveals the tough physical and neural tradeoffs that lie at the heart of each creature's anatomy, perceptual abilities, and behaviors. In the case of V1, foveal magnification makes sense because our eyes can swivel. In turn, we rely heavily on eye movements, day in and day out, because of that magnification. Our brains, like those of all creatures, invest in our strengths, specializing and adapting to leverage what we perceive or do well at the expense of what we do poorly.

How does magnification in a brain map affect what we perceive? It provides the neural manpower to represent *more stuff*. Often that extra stuff is fine-grained detail, like the flourishes of ornate filigree or the wending lines that form the letters on this page. The ability to perceive fine detail is called spatial acuity. Testing a person's spatial acuity typically involves presenting two things that are spaced close together and asking whether the person perceives them as one thing or two. For example, look at these parallel lines: ||. When you stare straight at them, you use your receptor-dense foveas and the expansive foveal representation in V1 to view the two lines. If your vision is normal, it will be obvious to you that the example shows two separate lines. But if you look at them out of the corner of your eye, you must make do with far lower acuity. Now, they will look like a single line . . . if you can make them out at all.

People with normal vision always have better visual acuity near the center of their visual field than in the periphery. But the details of how *much* better depends on the person. A pair of scientists set out to test whether these person-to-person differences in visual perception could be due to the idiosyncrasies of people's V1 brain maps. The team tested people's spatial acuity at different places in their visual field and also used functional MRI scans to see how each of their V1 maps were warped by magnification. They found that people differed in how much the foveal representation of their V1 map was magnified relative to the representation of the periphery. People also differed in how much better their spatial acuity was for details presented at their center of gaze than in their peripheral vision.

When the scientists compared people's foveal magnification in V1 and their changes in acuity across the visual field, they found that the two were linked. People whose V1 maps had greater foveal magnification also had the best spatial acuity for items viewed at the center of gaze, relative to items viewed in the periphery. Those with less dramatic V1 magnification showed less of a difference in their spatial acuity for items viewed at the center of gaze versus at the periphery. In other words, differences in how each person's V1 map is warped affect *what each person can perceive and where they can perceive it.*

It might take a moment to process what these findings actually show. What you can detect and perceive across the expanse of your visual field is different from what your best friend, sister, or neighbor can detect and perceive. Moreover, *how and where* you each detect well or poorly need not be determined by any failings of your eyes, but rather by the unique layout of your brain maps.

Scientific findings spanning more than a century have linked activity in the V1 visual map to what we consciously see. This does not mean that V1 independently constructs our conscious experience of vision; a wealth of evidence suggests that conscious experience arises from the coordinated activity of many different regions of the brain. Still, there is an undeniably strong link between what happens in V1 and what we consciously perceive. Consider the evidence thus far. The V1 map fills in the representation of the blind spot based on surrounding information, just as the white background of the page filled in the missing bird. Damage to the V1 visual map, like that suffered by Inouye's patients, creates zones of blindness in one's conscious visual experience. Both of these observations suggest a special link between the brain map, on one hand, and conscious perception on the other.

Observations like these are invaluable, but the most direct way to learn about this link is to intentionally muck around with a person's V1 visual map and then ask them what they see. It would be unethical, of course, to damage a person's brain in order to satisfy scientific curiosity. Luckily, neuroscientists are now blessed with ways of temporarily altering activity in a person's brain without damaging it. In a technique called transcranial magnetic stimulation, scientists apply a brief but powerful magnetic pulse at a spot on the surface of the scalp to make the neurons just beneath the skull more likely to fire. If I were to aim this instrument at the back of your head, centering the field on a part of your V1 map, you would see a point of light called a phosphene. The location of the phosphene in your visual field would correspond to the location in the V1 map that I stimulated. In other words, by momentarily zapping your brain map, I could make you consciously see a light that isn't really there.

A more invasive way of zapping the V1 map is to open up the skull and send electricity directly into the neurons within it. In 1968, Giles Brindley, a physician working on prosthetic devices for the blind, ambitiously attempted this. Brindley and his colleague placed an array of eighty radio receivers and eighty electrodes on the surface of V1 in a fifty-two-year-old patient who was blind in both eyes. When the scientists stimulated parts of her V1 map with the electrodes, the blind patient also saw phosphenes. Of course, due to her failing eyes, the woman would not have been able to perceive an actual spot of light. Yet by interfering with her V1 map, Brindley made the woman see light that did not exist.

Dr. Inouye's patients had perfectly good eyes and nonetheless were not conscious of seeing anything in their scotomas. Brindley's patient was conscious of seeing light even though her eyes no longer functioned. These are strange facts, leading to an even stranger takeaway: vision as you know it is born in the darkness at the back of your skull, reflecting what is happening in your visual brain maps more than what is happening in your two eyes.

That is why it matters so much exactly how your maps are warped: these maps, in turn, warp your conscious perception. This astonishing fact lies at the heart of this book and underpins the idea of a brainscape. A brainscape is the landscape that a brain map depicts. It is the distorted version of reality as it is mapped inside our heads. If the Boston subway map were a brain map, its brainscape would be the city of Boston stretched and distorted to proportions that matched those in the map. Likewise, your V1 brainscape is a version of the visual scene before you that is warped by magnification.

The illustration in Figure 10 attempts to render the V1 brainscape visible. Think about looking at the famous painting *Mona Lisa*. When you view this image, what you ultimately see is warped by your brain maps. In the figure, the thought bubble shows how an observer perceives more information and detail for the region of the painting at the center of gaze.

The resulting *Mona Lisa* in the thought bubble might strike you as wrong—of course you don't perceive a massive nose or lips every time you look at someone squarely. Instead, you glean a massive amount of information *about* the nose and lips: vital details for recognizing the person's identity and emotional state. Thankfully, your visual system has ways of correcting the effects of map magnification on the perceived size of objects. Imagine playing sports or using tools if you didn't make this correction. Objects would appear minuscule until you turned to look straight at them—at which point they would inflate to up to a hundred times their previous size. How would we ever grasp a coffee mug or catch a ball in such a world? Other areas of the brain help adjust your perception of object size, so that the mag-

FIGURE 10. A conceptual illustration of the warped V1 visual brainscape. The visual detail you perceive depends upon distortions in your V1 map—specifically, a magnified representation of visual detail at your center of gaze. ***Paul Kim***

nification in your V1 brainscape does not make objects appear to inflate and deflate every time you shift your gaze. This arrangement allows you to detect all the fine details of Mona Lisa's subdued smile without actually perceiving that her lips are as big as a room.

Despite this welcome arrangement, we still fall prey to some errors in judging size, which appear to stem from magnification in our brain maps. When psychologists actually measure people's judgments about the size of objects shown in different locations of the visual field, they find that our perception of object size *is* subtly influenced by where in our visual field the object is shown. Although we do quite well at judging the size of objects presented at our center of gaze, we consistently perceive objects presented in our visual periphery as smaller than they actually are. The farther out in your peripheral vision you see an object, the smaller you perceive that object to be. In short, magnification in your V1 map has a major impact on your perception of detail and might have a small effect on your perception of size as well.

PROBING OTHER BRAINSCAPES

Wilder Penfield was a pioneer of neurosurgery and neuroscience in the early twentieth century. He broke new ground in the treatment of patients who suffered from destructive epileptic seizures and, in the process, he became the first to chart the S1 touch map in the living human brain.

To treat seizures, Penfield took his scalpel right to the source: the brain itself. Patients were given a local anesthetic to numb the skull, allowing Penfield and his colleagues to open up the bony covering and reveal the serpentine folds of the cerebral cortex beneath. Before Penfield could excise the tumor or damaged tissue that was causing the seizures, he had to determine where that tissue was. He also had to determine the functions of the cortex surrounding it, so that his scalpel would spare tissue that the patients would need in order to feel, speak, and move for the rest of their lives. To do this, he inserted

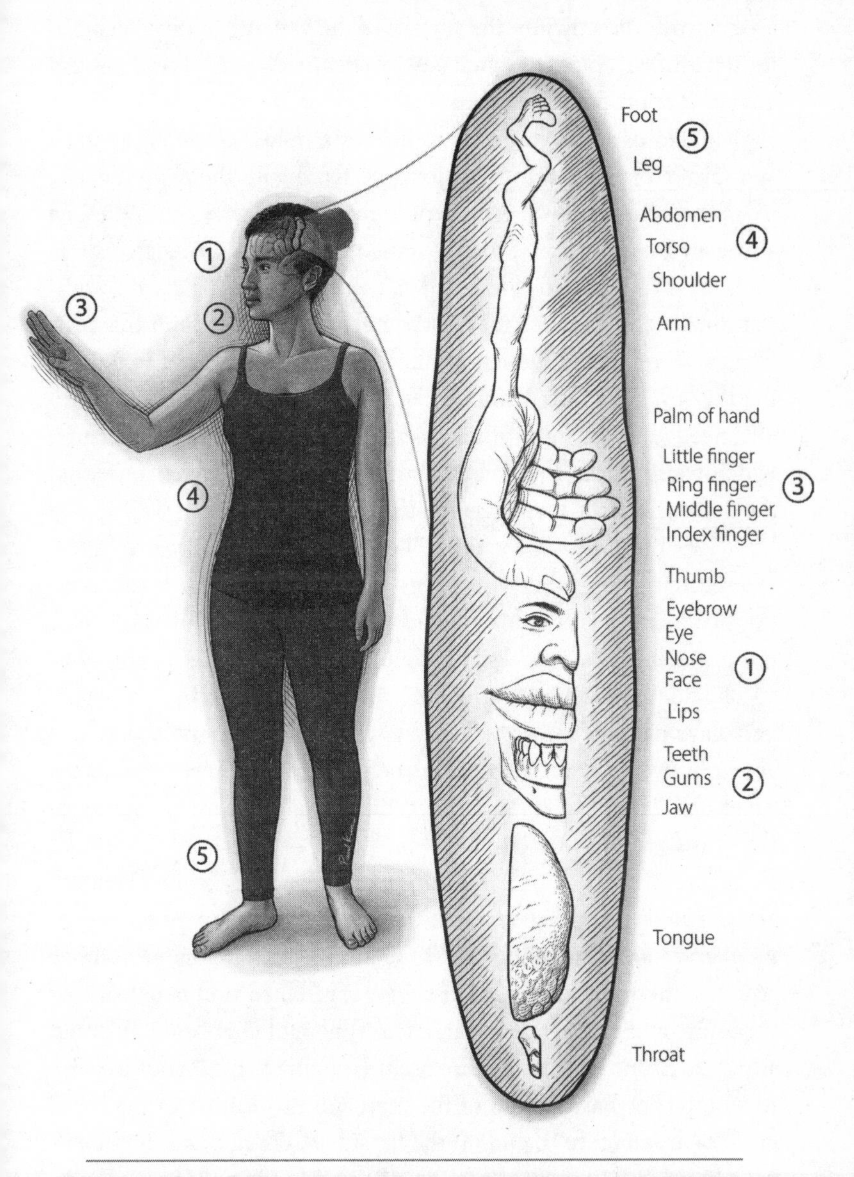

FIGURE 11. A schematic illustration of the human S1 touch map. The drawing shows half of the map, which represents touch on the opposite half of the body. ***Paul Kim***

an electrode directly into the surface of the patient's brain. The brain itself has no receptors to signal pain, so Penfield's patients did not feel any discomfort from the probe.

Penfield used his electrode to deliver a mild electric jolt to each spot probed in the brain. This jolt interfered with the natural activity of the neurons in ways that could reveal what they did. When he stimulated part of the S1 map, a patient might feel a tingle or some numbness in a part of the body. Patients were kept alert throughout the surgery because they had an active role to play in this process, from reporting the sensations that they experienced to following the surgeon's instructions to speak, read, or move on cue. They also needed to tell him if they started feeling the onset of a seizure, which would mean that Penfield's probing might have encountered the damaged tissue that triggered the attacks.

When the process was done, Penfield and his colleagues knew where the seizure-prone tissue was and which essential brain functions lay nearby. Using this knowledge, he could take out as much of the problem tissue as possible while sparing the patient's ability to speak and move. This technique of stimulating the brain during surgery gave patients the best chance of emerging from the experience able to move, talk, and function as before, but with fewer or even no more debilitating seizures. In fact, the technique was so successful that it is still widely used today.

In the process of probing hundreds of people's brains, Penfield and his colleagues learned about the layout of brain maps for movement and touch, including the S1 touch map. In humans, as in other animals, the right side of the body is represented on the left side of the brain, and vice versa. In each hemisphere of your brain, the map stretches from the side of your brain (roughly behind your ear) up to its top. The basic layout of the map and its location in the brain are shown in Figure 11. Out on the far side of the map, on the side of your brain, lies the representation of one side of your inner mouth, tongue, and lips. As you move inward and up, toward the top of the brain, the map represents the outer surfaces of that side of the face, then, for that side of the body, the thumb, the fingers, the rest of the hand, and at last the arm. Finally, beneath the tippy-top of your skull are the map representations of your torso, pelvis, leg, and foot on that side of your body.

The illustration shows something strange about the human S1 map: it appears to be scrambled, like a puzzle that was incorrectly assembled. Its grossest infraction is an abrupt transition from representing the forehead to representing the thumb, even though no functional construction of the human body could make such a leap.

The strange adjacency of face and thumb in your S1 map is an example of a discontinuity, or a point where the map breaks from an

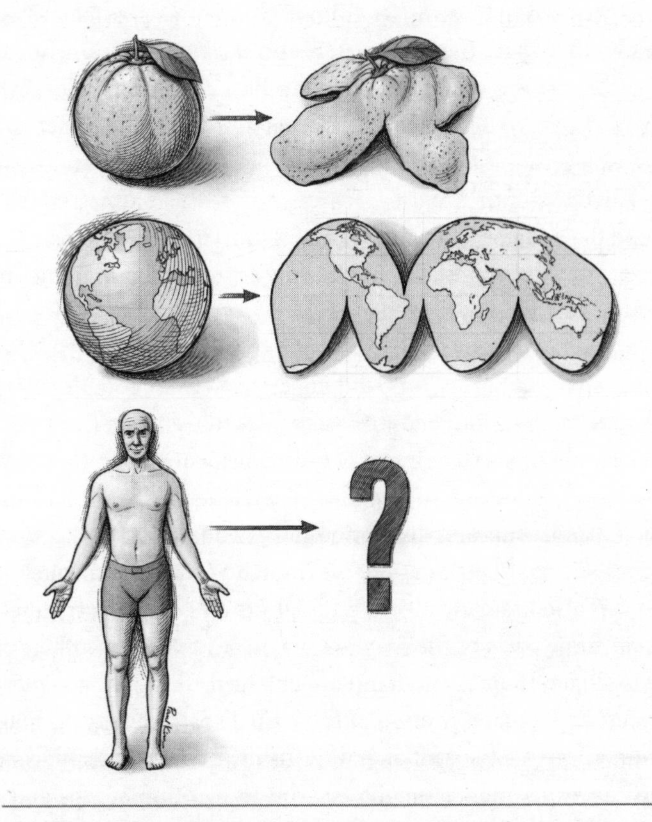

FIGURE 12. An illustration of the challenges inherent in creating two-dimensional maps of three-dimensional surfaces. *Paul Kim*

ideal, high-fidelity re-creation of your body's layout. At a discontinuity, the notion that neighboring inputs from the world (such as pressure on neighboring patches of your skin) are represented by neighboring bits of tissue in your brain falls apart. These lapses are small in most brain maps, but in certain cases, like the human S1 touch map, they can be dramatic. To understand these discontinuities, just think of the peel of an orange. (See Figure 12.) There is no way to flatten its spherical surface without cutting or tearing it. Mapmakers face the same problem when they make two-dimensional maps of the earth. You have to make a cut somewhere, ruining the sphere's continuity. If you read a world map literally, the eastern half of the Pacific Ocean and its western half are as far from each other as possible, whereas in reality they share the same waters and exchange the same waves.

To convert the surface of a sphere into a flat rectangle, cartographers also have to stretch out the parts of the globe closer to the poles, exaggerating the size of Europe, North America, and Australia relative to South America, Africa, and other landmasses that lie close to the equator. One kind of map—the Goode homolosine projection—avoids this pitfall by eschewing a rectangular map and making more cuts, as shown in Figure 12.

Of course, the human body is not a sphere. It features lanky protrusions (like your arms, legs, and fingers) and deep, complex caverns (like your inner mouth and throat). And so the challenges of cleanly transforming its surfaces into a two-dimensional map in the cerebral cortex are great indeed. As happens when flattening the peel of an orange, judicious cuts and discontinuities are unavoidable.

These discontinuities aside, the human S1 map that Penfield unearthed is also dramatically warped. Like the V1 map, the representations in some parts of the map are magnified relative to others. The human finger, thumb, and hand are enlarged, as are the tongue and face. Just as V1 magnification affects visual spatial acuity, S1 magnification affects tactile spatial acuity. To measure your tactile spatial acuity, an experimenter might ask you to feel tiny bumps and detect whether they are aligned or misaligned. Alternately, they might press points on your skin and ask whether you detect one touch or two. People have better spatial acuity on parts of their bodies that are magnified in the S1 map. For instance, you can probably distinguish between two pressures on the tip of your index finger if they are spaced about a millimeter apart, or less than the thickness of a dime. That distance would have to be seventy times greater—roughly the width of a woman's hand—for you to resolve them on your back.

These are just averages, of course. Like visual acuity, people's tactile acuity differs. The same scientists who studied people's visual acuity and V1 warping also tested people's tactile acuity on four fingers and used functional MRI scans to examine the warping of the finger region of the S1 touch map. They found that people who had more unequal acuity across the four fingers also had more unequally sized finger territories in S1, with the index fingertip getting the grandest estate and the pinky receiving a far smaller plot. People who demonstrated more equivalent acuity across their four fingertips had fingertip zones in S1 that were also more equally sized. In short, the results were the same for S1 as for V1: people's idiosyncratic abilities to perceive fine detail corresponded to the idiosyncrasies of magnification in their brain maps.

In fact, several parallels exist between V1 and S1. As Penfield's work revealed, stimulating neurons in S1 causes people to perceive tactile sensations when nothing is touching their skin. Distortions in the S1 map drive differences in how and where people can perceive fine detail with touch. People even exhibit errors in estimating size based on touch, just as they do with vision. Two prongs spaced a fixed distance apart will feel farther apart if you touch them with your index finger (which has a magnified S1 representation) than if you touch them with your forearm, thigh, or back. Here too, perceived distance is generally accurate when people feel with a body part graced with ample representation in their brain map. Yet they underestimate distance when using body parts that get short shrift in the S1 map. For vision and touch alike, perception as we know it is born inside our skulls. And for V1 and S1 alike, the limits of what we are capable of perceiving and the accuracy with which we perceive it is largely determined by the way our brain maps are warped.

These experiments beautifully demonstrate the basic relationship between magnification in the human brain map and our perceptual abilities. In fact, neuroscientists have predicted this connection for quite some time. This extraordinary link has profound implications. If what someone is capable of consciously detecting and verbally reporting is determined by the layout of representations within this person's brain maps, then we can hope to learn about their subjective perceptual experience by studying the objective, observable layouts of these brain maps. In essence, I can learn something about how it feels to be you by looking at your brain maps. Even if we do not speak the same language, or you do not speak any language at all, I can peer inside your head and gain real purchase on how you see and feel and sense.

But this ability to infer the subjective experience of others based on their brain maps is not limited to members of our species. If evolution were a poem, it would certainly rhyme; many motifs of brain and body organization resound across the animal kingdom, brain maps among them. An array of species, including all mammals, possess S1 maps that receive touch information from receptors in their skin in virtually the same way that your S1 map receives information from receptors in your skin. The same basic principles of touch representation and processing apply in the S1 of a rat as they do in yours. Of course, you have a different body layout than a rat does, and thus your skin surfaces have a different topography. That alone makes your S1 different from a rat's. But the starkest differences in S1 across different creatures stem from exactly how each map is warped.

Although each individual's S1 map is subtly warped in its own way, human S1 maps are far more alike than they are different, with fingers and lips grossly magnified relative to other regions of the body, such as the back and legs. The result is that we all generally have our best spatial acuity in the same places on our bodies—particularly our fingertips. Blessed with extra touch receptors in the skin and outsized representation in the S1 map, the human fingertip is to touch what the fovea is to vision. It is the human sweet spot for feeling things. But what about other creatures with other bodies and other ways of life?

How are their S1 maps warped, and what can that tell us about how that animal feels? Just as I can make some educated inferences about how it feels to be you by looking at your brain maps, we can do the same for a monkey, a rat, or a raccoon.

One of the pioneers in this undertaking was Edgar Adrian, a professor at the University of Cambridge who tackled the subject in the early 1940s. A lean and energetic man, Adrian was equally comfortable tinkering with an electrometer, facing down a fencing opponent, and blazing through quiet streets on his motorcycle. A decade prior, Adrian had won the Nobel Prize for his work demonstrating how neurons communicate with one another. Now, as political storm clouds gathered and erupted into the havoc of World War II, Adrian focused his scientific curiosity on the subject of touch. The skin sends signals about pressure on to the brain, but how? And what does the brain do with them? He tackled these questions in his laboratory, a dusty basement filled with antique scientific equipment that one visitor called "the most glorious clutter ever seen."

First, the cats arrived at the professor's laboratory. Then rabbits, dogs, and a few monkeys—nothing the university's School of Physiology hadn't seen before. When the sheep, goats, and pigs started coming, they surely drew some attention. By the time the Shetland ponies appeared, even the professor's dedicated assistant might have raised an eyebrow. But this motley assortment of creatures was eclectic by design. It was all part of Adrian's plan to unlock the secrets of their S1 maps.

Adrian had an electrical recording rig that allowed him to probe the brains of living creatures directly with an electrode. The setup was something like Penfield's, except that Adrian used his electrode to listen in on the ongoing activity of neurons rather than zap them to make them artificially active. Adrian's rig was attached to a loudspeaker, so that when his electrode detected blips of electrical activity from neurons firing, the loudspeaker signaled the event with a sound. By listening, Adrian found it easy to distinguish what the cells he probed were doing. When the electrode tapped neurons that were ready and waiting to receive information, but not signaling a sensation, he heard a "dull thud" from the cells. But when cells were excited, he heard what he called "a rushing noise" – a flurry of clicks indicating a rapid volley of neurons firing.

When the professor brought an animal – whether a cat, rabbit, dog, or monkey – into his laboratory, he put it to sleep with anesthesia so that it would remain still and not feel pain. Then he opened up a part of its skull and inserted an electrode, to listen in on the incoming touch signals arriving in S1. Each time he placed the electrode in a new spot, he would move around the animal, methodically touching it all over its body, and noting when he heard the rushing noise that meant the cells were excited. For each such noise, he noted where the electrode was in the brain and which areas of the skin, when touched, made those neurons fire. In this way, with his hands and his ears, his pen and his patience, Adrian could detect and reveal each animal's S1 touch map.

When it came time to study Shetland ponies, Adrian's assistant helped him arrange each anesthetized pony in a standing position supported by a wooden bench, its sleek head resting on sandbags. Once the animal was asleep and thus situated, Adrian excused his assistant and began the slow work of probing its touch map. He made his way over the animal's body again and again, touching the muzzle, the flank, the narrow pastern just above the hoof. He touched it with his hands, stroked it with a feather, or pressed its skin with a glass or wooden rod. It would have been a strangely intimate scene: the professor alone with the sleeping creature, prodding every surface of its body and listening to the music of its brain.

Adrian found that the nostrils reigned supreme in the S1 map of a pony. He wrote, "The area is divided into two parts of about equal size. The part in front is solely concerned with sensory messages from the area round the nostrils; the part behind deals with all the rest of the body surface – an area many thousand times as large as that of the nostrils." The pony's warped S1 touch brainscape is illustrated in Figure 13. The professor marveled at the blistering inequality of the pony's brain map and wondered at its significance. He noted that these animals bring their nostrils, and not their lips, into contact with fresh grass before deciding whether to consume it. Could the pony's nostrils be like our fingertips—a sweet spot for feeling through touch?

Among the creatures Adrian studied, he found dramatic warping in all of the S1 touch maps. But the specific *patterns* of warping—which parts of the body were magnified, and by how much—appeared to be uniquely tailored to the species that a given brain map served. For the sheep and goat, their prominent lips dominated the S1 map. In monkeys, the hands were magnified. In the dog and cat, the face, especially the whisker endings, was enlarged. The professor was particularly impressed with the pig's snout, which, as far as he could tell, occupied the animal's entire S1 touch map. He noted this body part's singular importance to the creature: "The pig's snout is its chief executive as well as its chief tactile organ, spade as well as hand, whereas the legs are little more than props for the body." Later studies revealed another region of the pig's S1 map, which Adrian had apparently missed because it lay hidden inside a fold of the brain. This hidden portion of the map represented the rest of the pig's body in its

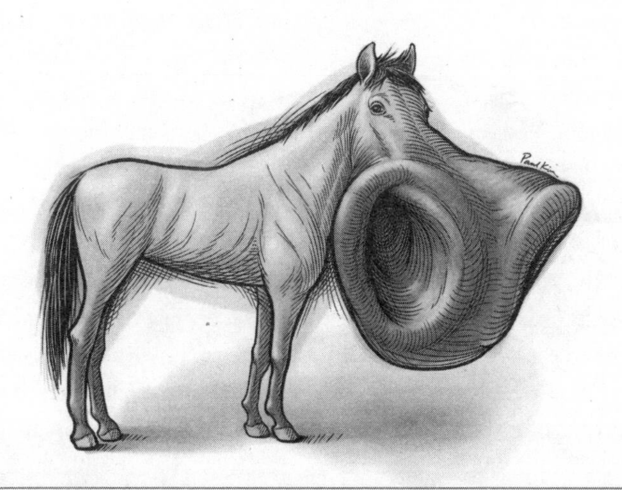

FIGURE 13. An illustration of the pony's S1 touch brainscape, showing its magnified nostrils. *Paul Kim*

entirety; it was only about half the size of the territory devoted to the animal's snout. (See Figure 14.)

Adrian observed one universal factor among all the species that he studied: some part of the head or face was magnified in the S1 touch map, presumably because, for four-legged creatures, the head and face stick out in front and are the first parts of the body to encounter new objects in the environment.

But what about monkeys and their fellow primates, humans? We rarely use our nostrils or lips to explore the world, yet we have faces and lips that are packed with extra touch receptors, and these features are represented by large zones in our S1 maps. Why is that? Adrian concluded that "the explanation is probably that we are descended from animals without hands, animals using the snout and the long [whiskers] of the face as their most delicate tactile guides." In other words, we feel more than we need to with our faces because our distant ancestors walked on four feet, like the pig, and survived better with faces packed with touch receptors. In this way, our perception of touch is shaped not just by our human bodies and human needs, but by the bodies and needs of the creatures from which we evolved.

FIGURE 14. An illustration of the pig's S1 touch brainscape, with its supremely magnified snout. *Paul Kim*

Since Adrian's early work, scientists have procured examples of warped S1 brain maps from a variety of exotic animals living under a variety of conditions. My personal favorite is the naked mole-rat, a relative of the rat that has carved out a life for itself underground in arid Africa. In the darkness of its burrows, the mole-rat knows the world by how it feels against its four protruding teeth. The mole-rat's incisors are its sweet spot for tactile exploration. When it comes upon an unidentified object in its burrow, it will tap its teeth against the object to find out what it is. Although the creature's four incisors combined make up no more than about 1 percent of the animal's total body surface, the representation of those teeth takes up nearly a third of its S1 map.

Another subterranean creature, the star-nosed mole, enjoys a damper habitat in North America. This creature has a wondrous star-shaped nose, with eleven tentacle-like appendages fanning out from

FIGURE 15. Photographs of the star-nosed mole (left), a close-up of one of its nostril stars with appendages labeled (top right), and a stained brain slice from the creature's S1 touch map, showing the corresponding representations of appendages in the map (bottom right). *From* PNAS, *vol. 109, supplement 1: "Evolution of Brains and Behavior for Optimal Foraging" by Kenneth C. Catania. Copyright © 2012 by the National Academy of Sciences.*

each of its nostrils. It uses this exquisitely sensitive nose to navigate dark, muddy tunnels and forage for worms, insects, and other food. When it finds something promising, it touches the item with one of those stubby appendages (called Appendage 11) that flank its mouth, to determine if the item is edible. The importance of the mole's spectacular touch organ is reflected in its S1 map: about half of it is devoted to representing the creature's tiny nose. But magnification doesn't stop there. The S1 map of the star-shaped nose represents all eleven appendages that surround each nostril, but it does not treat them equally. Although Appendage 11 is one of the smallest rays on each of the mole's stars, it claims five times the space in S1 than several of its larger appendages do. (See Figure 15.)

An obvious theme emerges from these warped maps. Brain maps in each creature become specialized to capitalize on the features the animal has in order to best meet its needs. The animal is born with certain inequalities in its skin; some zones are more conveniently placed for key tactile tasks and, thanks to evolution, these areas of skin tend to contain more touch receptors than others do. But just as your visual system takes a small inequality in the retina, between fovea and periphery, and magnifies that inequality a hundredfold in your visual brain maps, brain maps for touch build upon the inequalities in your skin to create even greater inequality in your brain.

These inequalities guide our actions and choices. If you want to know whether a fabric is cotton or polyester, do you feel it with your elbow or with your hand? If you want to know whether a pear is ripe or lettuce is crisp, do you press it with your knuckles or the pads of your fingertips? Reaching to touch with your fingertips is just like shifting your gaze to place objects of interest on your visual sweet spot, the fovea. Your warped brainscapes spur you to move your eyes to look or to extend your hands to touch because *you can feel, see, and discern so much more* when using these perceptual sweet spots. Thanks to inequality and magnification, your fingertips allow you to tell cotton from polyester, just like the star-nosed mole's Appendage 11 allows it to discriminate between a worm and an inedible stone. From the lips of the sheep to the nostrils of the pony and the teeth of the mole-rat, inequalities provide all creatures with sweet spots for discerning the world around them through touch.

Since brain maps operate under the same basic principles across species, they can be used as guides to relate our own perceptual experiences to those of other animals. There isn't an exact calculus for transposing one creature's perceptual experience to that of another. But Adrian took a stab at it when he was studying hoofed animals in his basement laboratory. He wanted to get a handle on just how impressive the pig's snout is relative to the human hand. He compared the physical surface area of the pig's snout to the surface area of its snout's representation in its S1 brain map. For every square centimeter of cerebral cortex that represented the animal's snout, there were about 10 square centimeters of skin on the surface of the snout itself: a ratio of 1 to 10. Then he worked out the numbers for the human hand. For every square centimeter of cerebral cortex representing the human hand, there were 75 square centimeters of hand surface: a ratio of 1 to 75. These rough numbers suggested that the pig can feel *more* fine detail with its snout than you can feel with your hand.

Armed with the pig's S1 brainscape, the human's S1 brainscape, and your perceptual experience, it is possible to make some reasonable inferences about what it feels like to be a pig. When the pant leg of a farmer brushes against the face of the pig, it wouldn't feel like what you feel when fabric brushes against your face. It would feel like *more*. And the best you can do at approximating that feeling is to imagine that you had your hand, rather than your nose, growing out of the middle of your face, with your palm open to the fabric and the world. You would feel the weave of the fabric, the path of its stitching, the fabric's warmth, and its weight. And if your porcine brain had any inkling of such things, you would know in an instant if the slacks were made from polyester or cotton.

If this is your first time imagining how it feels to be a pig, you have almost certainly never imagined life as a rat. Whiskers are to a rat what the snout is to a pig. The skin at the base of each whisker is exquisitely sensitive to any slight deflection of these stiff, specialized hairs. Whiskers are an ingenious invention because they allow animals to extend their sense of touch outward, beyond the face. The nocturnal rat can navigate sewers, fields, or basements quickly in the dark because of the whiskers that stick out in front and to the sides of its muzzle. As the animal approaches objects or obstacles in the darkness, the whiskers are disturbed, telling the rat where things are in its immediate environment. Using their facial muscles, rats can also sweep their whiskers back and forth in long, regular strokes that allow them to collect different types of tactile information, including details about the shape, texture, and identity of objects in their path.

The whisker domain of the rat's S1 touch map dwarfs everything else, taking up at least a fourth of the overall map. This whisker area contains zones, and each zone represents the deflections of a single whisker. As Figure 16 shows, the layout of these zones in the S1 map

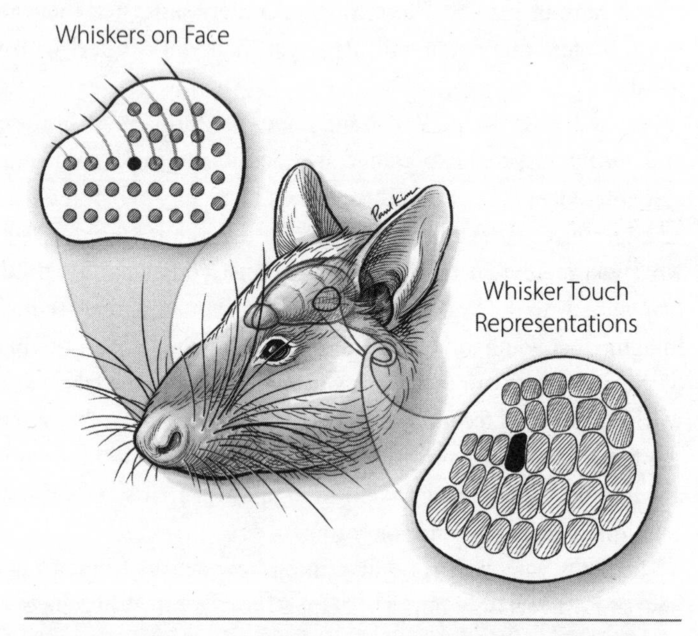

FIGURE 16. The layout of whisker representation in the rat S1 touch map reflects the layout of whiskers on the creature's face. ***Paul Kim***

beautifully mirrors the layout of whiskers on the animal's face. The illustration in the first panel of Figure 17 shows how the rat's S1 touch brainscape is dominated by its representation of the whiskers. The second panel is a photograph of a rat's brain stained to render its S1 touch map visible.

We don't have rodent-like whiskers, but our fingernails and hair are in many ways similar to them. Like whiskers, they don't contain touch receptors, which is why you can trim your nails or hair without feeling pain. But you can certainly feel with your scalp when your hair is pulled or feel with the bed of your fingernail if that nail is lifted or torn. Rats probably experience movement of their whiskers in a similar way, but with far greater acuity and sensitivity. Imagine that you have long, flexible nails growing out of the pads of your fingertips. As you move your hands around, those nails bend and deflect, which tugs your fingertips in different directions and to various degrees. If those fingertips were embedded in the flesh around your

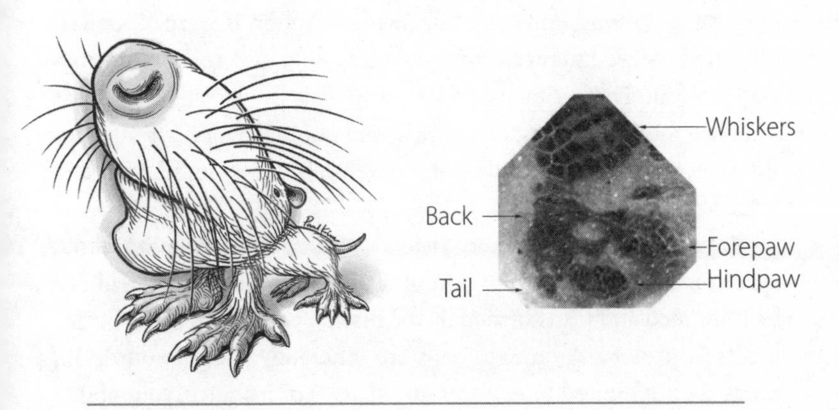

FIGURE 17. **The rat's S1 touch brainscape (left) reveals the outsized importance of its whiskers and snout. A photograph of a stained brain slice from a rat's S1 touch map (right) shows the layout of its body map, including its tail, paws, back, and some of its whiskers.** *Illustration by Paul Kim. Photo from* Current Biology, *vol. 26, no. 1. Copyright © 2016 by Elsevier Inc.*

nose rather than swinging about at the ends of your arms, you would have some idea of how having whiskers feels to a rat.

This exercise in rodent embodiment takes on a new dimension when you consider how rats use their whiskers to interact with other rats. You could call it the rat version of a handshake, but that might be understating its importance and intimacy. In the process of meeting or greeting, two animals line up, nose to nose, so that their noses touch and whiskers overlap. Instead of moving their whiskers in long, regular sweeps, the rats make small, irregular whisker movements. They are, in a sense, twiddling each other's whiskers, which intensely stimulates the most sensitive touch organs of both rats. During this social touching, neurons in the whisker region of the rat S1 map fire at especially high rates. The rat's warped S1 map and the activity of the neurons within it provide a bridge between the rat's humble behavior and its sensations, giving us an inkling of the intensity experienced during this simple social act.

Brain maps shape our perception of our world in powerful ways. Together with the brainscapes they represent, brain maps have a lot to tell us about why we feel and see and act the way we do. They render visible the ways in which our own perception is warped, underpinning how we interact with the wider world and collect information from it. They offer us a view of what is universal and what is unique about our senses, our minds, and our behaviors. They bridge the objectively visible and the subjectively felt, across individuals and even across species.

Brain maps of touch and vision show how information about space on the surfaces of the skin or space within the visual field can be translated and transformed by the brain. But that is just the beginning of the story. Although maps are inherently spatial, your brain maps are not limited to representing space. For instance, your ability to cull information from sound depends on the invisible spectrum of frequency. As you will see, you are privy to this spectrum only because of a brilliant anatomy and the wonders of brain maps.

4

Out of the Ether: Brain Maps for Hearing

ONE MORNING IN THE 1940S, six-year-old Gerald Shea awoke in his home north of Manhattan feeling uncharacteristically tired. The family doctor visited and discovered that the boy had both chicken pox and scarlet fever. For two brutal weeks, Gerald weathered an onslaught of rashes, fever, and fatigue. It took several more weeks for his pocked skin to heal and his scars to fade. But the lasting damage from the illness was invisible to Gerald, his family, and even his doctor. It happened in the deep recesses of his inner ears, where infection had destroyed precious cells in each cochlea, leaving him deaf to high-frequency sounds.

Young Gerald did not consciously perceive this deafness even as his world was transformed by it. He no longer heard the rustle of wind blowing through trees, the tapping of raindrops on a rooftop, or the reassuring sound of his own footsteps. Most important, he lost the ability to hear many speech sounds. Although he was still able to carry on conversations, comprehending other people's speech had become effortful and would prove to be a lifelong struggle.

Gerald's challenges demonstrate the structure and surprising complexity of natural sounds. Although we don't consciously perceive this structure, we use its information to make sense of the world around us. When Gerald lost the ability to hear high-frequency sounds, those created by certain objects dropped away completely. If he listened to a symphony or an aria, the flutes, violins, and soprano vocals would be inaudible, while the cellos, tubas, and other lower-pitched instruments continued to transmit sounds that he could hear. In a conversation, consonant sounds like *s* and *g* and *t* vanished, while vowel sounds like *a* and *o* remained. Different sound frequencies transmit different types of information.

Sound frequency lies at the heart of how we use and make sense of sound. Yet unlike an object's location in visual space or a touch on the body's surface, frequency is not a spatial phenomenon. Can frequency be represented by inherently spatial maps? Absolutely. In fact, auditory maps demonstrate the versatility of brain maps.

Sound begins with a physical event, some form of movement that causes one or more objects to vibrate. The physical event could be any number of things: the collision between the sole of your shoes and the surface of the floor, the rending apart of fibers as a sheet of paper tears, or helicopter blades slicing through the air. But in all cases, the result is vibration. First, the objects involved in the physical event begin to vibrate. Then their vibration jostles the surrounding air molecules, causing that vibration to spread from the object to the air molecules around it. These air molecules jostle *their* neighboring air molecules, and so on, forming a pressure wave. In this way, the vibration that started in an object is now transmitted outward from the object through the air.

There are two important things to know about any given vibration. The first is its amplitude. The amplitude is *how much* the vibrating thing is displaced by the vibration. It corresponds to how loud a noise will sound to us. High-energy events like a gunshot create vibrations that displace air a lot, and therefore they sound quite loud if they happen near us. As the vibration travels outward from the event in all directions, it loses energy. That is why sound can tell us about high-energy events when they happen near or far away, but it can tell us about low-energy events only if they happen nearby. You have understood this concept since early childhood, although perhaps in different terms. It is why you would shout to speak with a playmate across the street and whisper to share secrets with a classmate beside you.

The other thing to know about a vibration is its frequency. The frequency is *how many times* something (such as an object or air molecules) moves back and forth in one second. Frequency is measured in hertz, or cycles per second. From frequency, we get our perception of a sound's pitch, or whether it sounds high or low to us. But that is just one small part of what we glean from frequency, because all vibrations in the natural world are complex and messy. Instead of having a single frequency, they are made up of many different frequencies all at once.

Whether it is the wail of a saxophone solo or the drip of a leaky faucet, sounds are richly complex and comprise a wealth of frequencies. The profile of frequencies that make up a sound contains vital information about what that sound is. The best way for a physicist or an acoustical engineer or your brain to characterize a sound is to break it down into the profile of individual frequencies that it contains. This profile is the hidden structure within sound. Our brains use it to determine a sound's timbre, which is essentially how something sounds or what it sounds like. Thanks to timbre, a trumpet and a violin can play the same musical note and still sound entirely different. Thanks to timbre, you can recognize a familiar voice—and understand what it is saying besides.

Thankfully, our bodies and brains are superbly designed to capture the hidden structure in sounds and to capitalize on it. This clever design begins in the ears, which contain the crucial machinery that makes hearing possible. The human ear contains several miniature bones and membranes. Operating together, they are a marvel of engineering achieved over millennia of trial and error. Every part of the system contributes in some way to the task of gathering sound.

But it is within the tiny, coiled cochlea where the true business of hearing takes place. Here, incoming vibrations are translated into the language of the brain, so that we can ultimately experience them as sounds.

The cochlea is curved into a spiral. It is tiny—no larger than a baby pea—yet it is our gateway to the world of sound. Uncoiling the cochlea would be as difficult as unfurling a snail's shell, but if you could do it, you would be left with a fluid-filled tube about 35 millimeters long. An intricate architecture of tissues and cells exists in that watery tunnel, but its most essential feature is the rows of bristles that stretch along its length. These bristles are the tips of delicate sound receptors.

When the pressure wave from a nearby event reaches your ear, the tiny machinery within gets to work amplifying the wave and sending it into the watery cochlea, where the wave continues on through fluid instead of air. This wave disturbs the miniature architecture of the tube and makes the tiny bristles of your sound receptors move. This movement makes the receptors increase their firing rate, sending off a signal to the brain that a sound has been detected. But these receptors have a secret organization. They are not a helter-skelter crowd of fourteen thousand random cells. Their orderly rows run the length of the cochlea's tube, organized according to the landscape of frequency. The receptors that rest at one end of the tube, making up the outer coils, are excited by the high-frequency components of a vibration. Those at the other end, near the center of the cochlea, are excited by low-frequency components. If you unfurled the cochlea and made your way slowly from the former end of the tube to the latter, you would encounter row after row of receptors, with each set tuned to detect lower frequencies than the ones before.

The beauty of the cochlea and its receptors is that they use the laws of physics to automatically deconstruct every complex natural sound into all of the simple frequencies that comprise it. In the process, they take frequency, a phenomenon related to *time,* and convert it to a representation based on *space,* in this case, the one-dimensional space that runs the length of the cochlear tube. The activity of receptors at one end of the tube represents high-frequency sounds, whereas the activity of cells at the other end of the tube represents low-frequency sounds. When the receptors send their signals on toward the brain, this spatial representation of frequency is preserved. Their messages travel through several stations buried deep within the brain, which preserve this spatial code, before arriving at the primary auditory cortex, or A1, in the cerebral cortex.

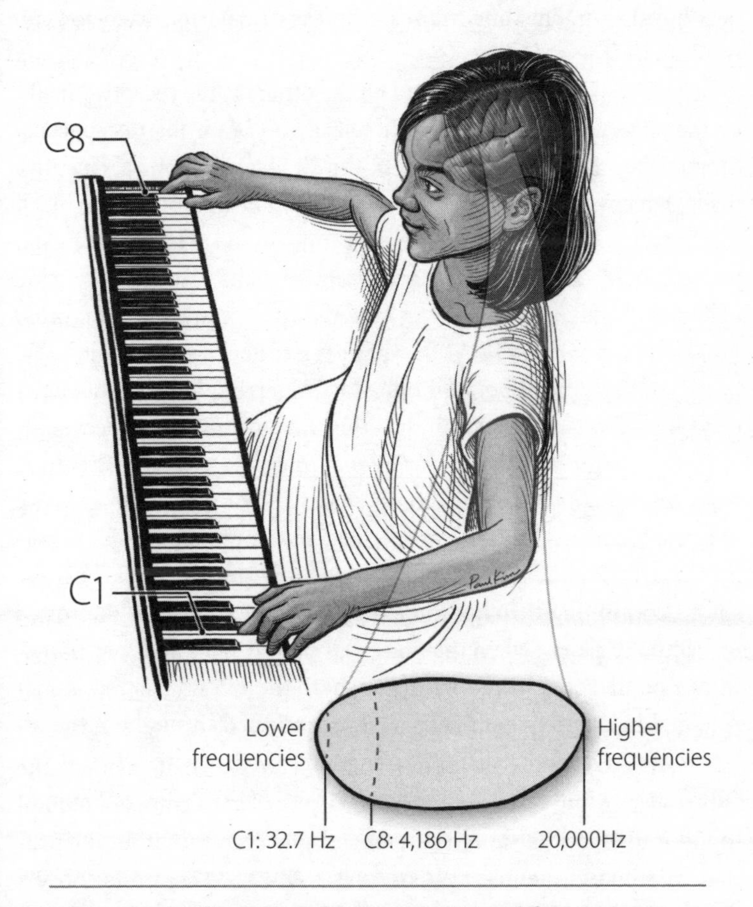

FIGURE 18. An illustration of the human A1 sound-frequency map.
Paul Kim

Just as your primary visual cortex contains a map of your retinas (and thus your visual field) and your primary somatosensory cortex contains a map of your skin's surfaces (and thus your tactile inputs), your primary auditory cortex represents sound using a map of your cochleas. The illustration in Figure 18 shows how A1 is laid out in a continuous map of simple component frequencies, with lower frequencies represented at one end and higher frequencies represented at the other. This brain map undergirds your conscious perception of sound. Stimulating this region with an electrode causes a person to hear buzzing or whistling sounds, whereas damage to this region can result in deafness.

The key to the map in A1 and all the others is the receptive fields, or the slivers of reality that each cell in the brain map is specially tuned to represent. The neurons in the V1 visual map have receptive fields that cover a portion of the visual field. The neurons in the touch map have receptive fields that cover a zone on the skin's surface. And the cells in the auditory map have receptive fields that span a portion of sound-frequency space – say, sounds with frequencies of around 1,000 hertz. A neuron in A1 that prefers frequencies of around 1,000 hertz will have neighbors on one side preferring lower frequencies (say, 900 hertz), and those on the other side will prefer higher.

The benefits of having a frequency map in A1 are no different from the benefits of having a body map in S1 or a visual map in V1. Like the V1 map, A1 fills in unexplained and improbable holes in perception. If a loud noise interrupts an ongoing sound, people hear the sound continuing through the loud interruption, even if the sound was actually absent when the noise was played. This auditory filling-in can be detected in the A1 map, where the missing sound is represented in the frequency map as if it continued throughout the interruption. Thanks to this filling-in, you can chat with a friend at a coffee shop without having to start over every time someone coughs or the grinder whirs.

Like other brain maps, the A1 map supports local processing, but in this case local processing means comparison between similar frequencies rather than between similar points in visual or tactile space. In A1, neurons tuned to similar frequencies can be heavily interconnected with short little wires, saving precious energy and space in the brain. Local processing in A1 helps your auditory system identify key frequency structures that make up complex sounds. This, in turn, helps you determine what those sounds are.

To understand how essential this process truly is, look no further than human speech. When you speak, you create vibrations by forcing air through the vocal folds in your larynx. You use the resonant properties of your throat and mouth, and the moment-to-moment placement of your tongue, lips, and teeth, to craft the specific complex sounds that leave your mouth.

If I were to mention to you the saying *Easy come, easy go,* the vibrations leaving my mouth would look like those shown in the graph in Figure 19. This graph represents frequencies using space. It shows the component frequencies within sounds vertically, so that lower frequencies are depicted at the bottom of the image, while higher frequencies are depicted near the top. Along the horizontal axis is time, showing how the complex sounds would unfold as I spoke. The darker a point in the image is, the greater the amplitude of that component frequency in the spoken sound at that moment in time. If you were actually listening to me say these words, the vertical patterns of dark and light in the graph would be re-created in the activity of sound receptors in your cochlea and the neurons in your A1 sound-frequency map. Local processing in A1 would help your brain detect the crucial underlying structure and convert this sonic mess into an effortless understanding of my spoken words.

The dark horizontal bands that stand out in the graph are called formants. They are found in all vowel sounds, which are essentially speech sounds produced when your mouth is mostly open and the air can flow more or less freely through it. When you produce an *eye* sound and then an *oh,* your mouth and tongue will take different positions, changing the way vibrations from your throat and mouth resonate. Every vowel has its own unique formants, positioned at differ-

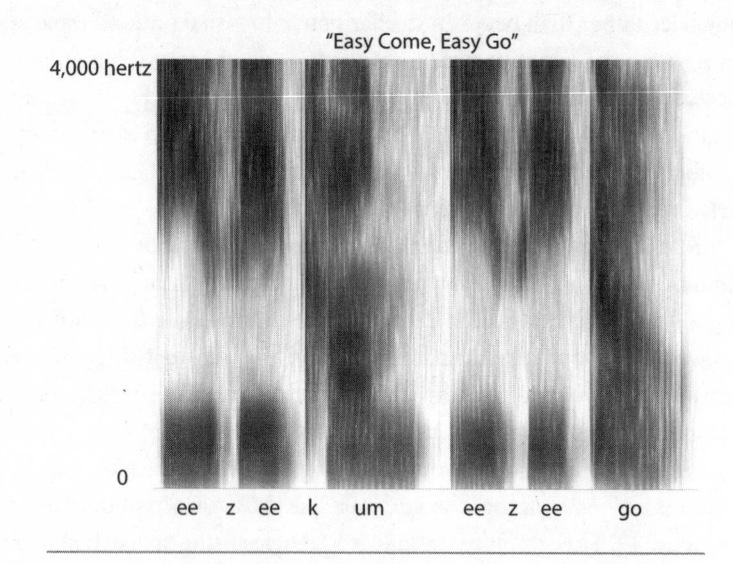

FIGURE 19. A graph showing the frequencies that make up my voice as I say, "Easy come, easy go."

ent frequencies. We use the position of the lowest three formants to determine which vowel sound we are hearing. The graph in Figure 20 depicts the frequencies I emitted while saying the words *eyes, as,* and *owes*—three words that differ only in vowel sound. The lowest three formants of each vowel are marked with arrows. Different languages have different vowel sounds, with their own unique formant spacing. But, thanks to the similar throats and mouths of humans around the world, the vowels in all spoken human languages resemble those in the picture and are identified by their formants.

The speech sounds for consonants come in several flavors, each with its own unique acoustic characteristics. Some consonants momentarily stop the flow of air and then release it in a *burst,* like both the *b* and *t* sounds in that word. Others consonant sounds, like *s,* are made by forcing air through a small opening in the mouth. This creates air turbulence, which yields a satisfying high-frequency hiss.

The clues to determining which speech sounds, and thus which

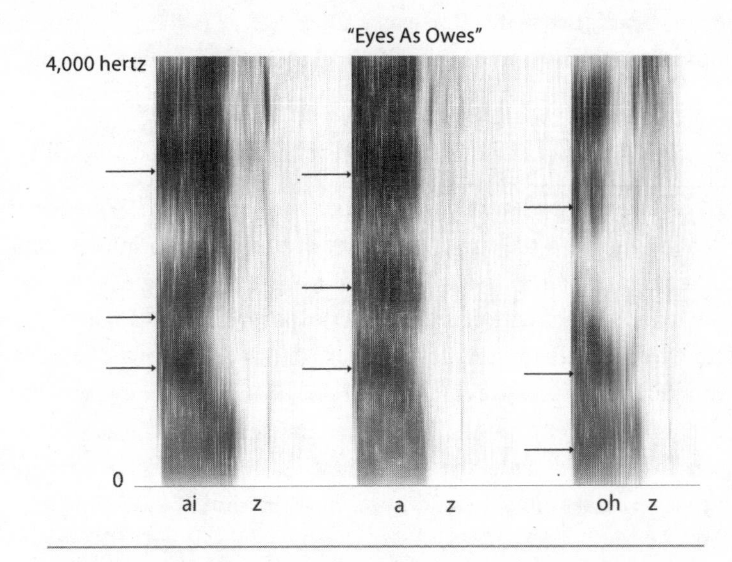

FIGURE 20. A graph showing how vowel identity changes with formant (dark band) spacing for the spoken words *eyes, as,* and *owes.* Formants are indicated with arrows.

words, I am saying lie buried in the frequencies that make up those sounds. That is why Gerald Shea struggled with speech comprehension. When he lost his hearing for high frequencies, he was left with incomplete information about the underlying structure of sounds. For many speech sounds, including most spoken consonants, he simply could not hear the frequencies that define what that sound *is.* By paying close attention to people's lips and treating every incomplete utterance like a puzzle that he had to solve, he was able to fill in many of the gaps and carry on conversations. But the process was effortful and not without its errors. Someone saying "Her way of speaking gently" might come across as "Arrays of seas he lent me." Or a clause like "What'll happen after Nora leaves" might be deciphered as "Water happens after coral reefs." In spoken language, the common currency of communication for most people on earth, sound frequencies are the bridge between what one person means and what another person hears. Every day, for every utterance, it is the information transmitted through frequency that makes comprehension possible.

OTHER WAYS OF HEARING

Just as brain maps let us peer into our own sense of hearing, they offer insights into the ways that other creatures wring information out of vibrating air. Human hearing at its best (in young people who have avoided loud concerts and other exposures to loud noise) can detect frequencies ranging from about 20 to 20,000 hertz. That may sound like an impressive range, but it does not hold a candle to what many of the other creatures that roam and scurry on this earth can hear. Take, for example, the humble rat. As you now know, rats are intensely social animals. In addition to their whisker-twiddling version of a handshake, they also socialize by using whistle-like vocalizations at frequencies above 30,000 hertz—far too high for human ears to detect. Adult rats chat with new acquaintances for short bursts at frequencies of up to 60,000 hertz, whereas rat pups let out longer, plaintive calls to their mother at frequencies of up to 40,000 hertz.

The rat's A1 sound-frequency map reflects this range, transitioning smoothly from neurons that prefer frequencies as low as 1,000 hertz at one end to those preferring frequencies of up to about 70,000 hertz at the other. When a rat hears social vocalizations comprised of high frequencies, these sounds are represented in the high-frequency end of its A1 map. Just as the pattern of activity across your A1 map reflects a landscape of overheard frequencies, so too does the activity in a rat's A1 map. But thanks to your auditory differences, those landscapes will be quite different, even if you are in the very same room. Some quiet evening at home, you might pause to think of the rat greetings that could be taking place near you though literally falling on your deaf ears.

Although frequency is the mother tongue of hearing and the primary dimension that determines auditory brain maps, it is by no means the only dimension. Likewise, we do not just use sound to tell us *what* something is, but also to tell us *where* it is. Our flying mammalian brother, the bat, provides a beautiful demonstration of how brain maps can represent *where* a source of sound is located. As nocturnal predators, bats rely on their ability to navigate and catch prey in darkness. Instead of waiting for prey to let out a noise, bats emit their own complex sonar pulses and then locate prey in darkest night by analyzing the echoes that return to their ears.

For an example, consider the mustache bat, so-called because of the dapper ring of long hairs that grows around its mouth. The mustache bat lives in large colonies in forests or arid regions and hunts insects at night. There are three phases to this hunting process. First, the bat needs to detect prey in its environment (the search phase). Then, once it detects something, it swoops rapidly toward the target (the approach phase), before closing in on the prey at close range (the terminal phase). Using its sonic pulses and their echoes alone, it can detect a small fruit fly from more than 3 meters away and gauge its prey's speed with an accuracy of about 10 centimeters per second —all in the dark. When these bats are homing in on prey, they judge their distance from the target by how long it takes for their calls to echo back to them; the later the echo, the farther they are from their prey. To the bat's brain, the echo delay stands in for distance. Put another way, time equates to space.

Mustache bats have a special region of the auditory cortex that processes echo delays. Neurons in this part of the brain will fire in response to an echo, but only if a specific amount of time has elapsed between the bat's call and its echo. This part of the cortex contains a continuous map of echo delays, with neurons that respond to the shortest delays (less than 0.5 milliseconds) at one end and those that prefer the longest (around 18 milliseconds) at the other. The illustration in Figure 21 shows the echo-delay map, with dotted lines and labels indicating landmarks within the continuous map. Although I've described and labeled the map using units of time (milliseconds), the map is ultimately one of space: specifically, the bat's nighttime, midflight hunting environment. This space begins a few centimeters past the creature's own mustached snout and stretches several meters into the dark.

This time-space echo map also stands as another example of how brain maps may be warped by magnification to serve an animal's needs. When echo delays are between 3 milliseconds and 8 milliseconds, this means that the prey is between 50 to 140 centimeters away from the bat. This corresponds to the approach phase, when the bat

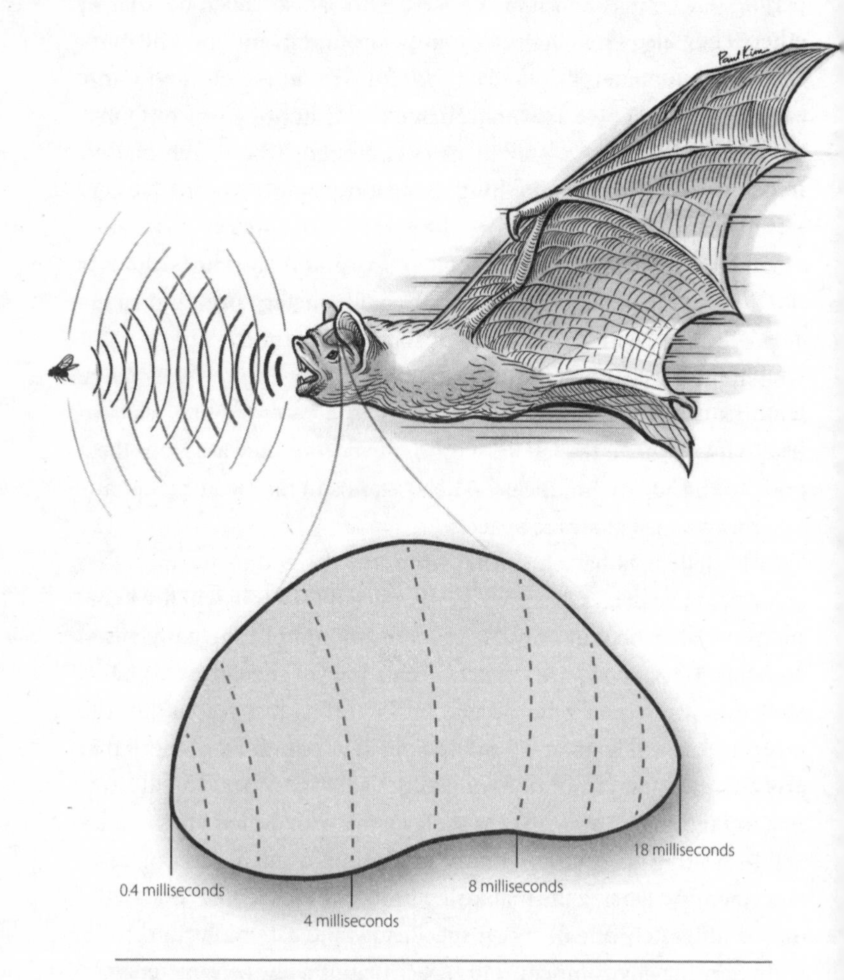

FIGURE 21. An illustration of the mustache bat's echo-delay map, in which echo delay (time) equates to a spatial dimension: the bat's distance from its prey. ***Paul Kim***

swoops toward its target. Neurons in the echo-delay map that prefer delays in this range take up a disproportionately large swath of the map, magnifying and prioritizing information to guide the bat's swooping approach.

Auditory brain maps reveal the complexity of sound and the surprising ways in which we coax vital information from the vibrating ether. They also demonstrate how spatial maps can represent nonspatial phenomena. This is the beauty of representation. Just as you can draw a map of the Egyptian pyramids on almost anything (foggy window and sandy beach alike), you can build maps of distance, time, frequency, temperature, and more, all with the same set of materials – in the case of brain maps, neurons, electricity, and time.

Dimensions of distance across the skin's surface, dimensions of time to echo delays, and dimensions of vibration frequency are all continuous dimensions. Any given temperature relates to any other temperature in a clear and mathematical way. Ultimately, the two temperatures are merely two points on a single dimension of thermal heat. The same can be said of frequencies, distances, or delays. Brains represent many such dimensions with maps.

Yet some aspects of our world are better characterized by categories than by dimensions. How do you relate the taste of brown sugar to that of mustard, or the scent of pine needles to that of talcum powder? The meaningful stuff in our world defies simple definition and eschews singular dimensions. Do brains make maps for types of stuff? The two chemical senses, taste and smell, are both tasked with categorizing and identifying substances. Through their neural representations, these senses reveal both the extreme versatility of brain maps and their severe limitations.

5

Brain Maps and Codes for Taste and Smell

To SURVIVE AND MULTIPLY ON EARTH, all creatures must eat stuff, mate with stuff, and keep away from dangerous stuff. When you come down to it, life is really about *stuff.* And what is important to a creature is important to its brain.

Your sense of taste and sense of smell are tasked with determining what kind of stuff is in your mouth or near your nose, respectively. A closer look at these two senses and their representation in the brain reveals the specific and often unexpected ways in which they protect both animals and humans. These senses also demonstrate the limits to what brain maps can represent and show how something else—a neural code—helps carry the load when brain maps cannot.

A BODY UNDER SIEGE

It is easy to think of taste as a vehicle for experiencing pleasure. Each mealtime can be an opportunity to enjoy comforting, exciting, or indulgent flavors. But the basic purpose of taste is not to bring you pleasure. It is to keep you from dying.

Taste is important because, at the end of the day, you are a glorified sack of chemicals. Everyone is. When it comes to survival, what matters is the division between the vital stuff within your body and the myriad hazards outside it. Your body must be stocked with all of the necessary molecular building blocks and vital fuels required to keep you alive and yet free of common toxins that would do you harm. Although skin has a role in absorbing certain compounds, largely it acts as a fortress wall, surrounding you and keeping out intruders. Any city under siege faces a critical challenge: how to keep the enemy out while letting in food, water, and other supplies. Your body faces this challenge every day of your life.

All that you are, and all that allows you to continue to be, depends on the chemicals within you. Ingest the wrong compound—something poisonous, rotten, contaminated—and you will cease to be alive. Fail to ingest the right compounds day in and day out, and you will cease to be alive. Our world is filled with a vast array of compounds that you, a mouse, or a fruit fly might try to nibble. Thankfully, we don't need to understand the difference between an amino acid and an alkaloid in order to choose between them at the table or the trough. Nature, or rather evolution, has given us a cheat sheet—a practical guide for ensuring that the right stuff goes in and the wrong stuff stays out.

Those who study the chemical senses distinguish between taste and flavor, terms that we tend to use interchangeably. A food's flavor is the combined experience of its taste, its smell, and even its texture. The odors that a food gives off, its temperature, and its feel on the tongue all add a great deal of subtlety to the experience of its flavor. If you have ever tried to enjoy a fine meal despite a stuffy nose, you will have some idea of how very different flavor and taste can be.

The sense of taste begins with taste receptors that coat the tongue and other surfaces in the mouth. Taste receptors come in several different varieties, and each recognizes different types of chemical compounds in food. Each type of taste receptor is linked to one of two innate, or hardwired, reactions: swallow the food (and shovel more in) *or* eject the food from your mouth. Attraction or repulsion. In or out. Yes or no. Your entire sense of taste, and many hundreds of thousands of taste receptors bundled into thousands of taste buds, comes down to that one binary choice.

It might be surprising to learn that, for all the pleasure we derive from eating, only three types of taste receptors are known to drive our attraction to food. Sweet-taste receptors detect sugars and other carbohydrates—critical sources of energy for the brain and the rest of the body. They are sometimes fooled by other molecules that resemble sugars, which is why artificial sweeteners taste sweet to people even though they offer no nutrients or fuel. The umami (savory-taste) receptor detects free amino acids that indicate a food is high in protein. Amino acids are the building blocks of proteins, which are in turn the raw materials for pretty much everything that makes complex organisms function. Although your body can often recycle amino acids, it can't manufacture many of them. So humans and other animals need to get these building blocks by eating foods that contain protein.

Saltiness reveals how delicate the balance of chemicals in your body truly is. Life on earth evolved in the sea, under conditions in which sodium salts (sea salt, or NaCl) were easy to come by. As a result, basic cellular functions supporting life on this planet depend on salt. In order for our bodies to function, we must all be a little bit salty inside. That is why hospitals hydrate patients using intravenous saline solutions that are, in essence, water mixed with a pinch of sodium salt. It is also why many other land-dwelling animals go to great lengths to ingest salt; in order for their bodies to function properly, they need to keep their insides salty too. Yet ingesting too much salt can be a problem. Overindulge in sodium, and over time, you could be on the path to problems with blood pressure and kidney function. And if your insides were to suddenly become extremely salty, you would have a medical emergency on your hands.

To survive this delicate balancing act, you possess two types of taste receptors for salt. One responds to low levels of sodium salts—the good kind of salt your body needs—and triggers an attractive response. Snack foods and restaurant fare tend to be salty because, at that level of saltiness, your attractive salt receptors send signals to your brain that make the food taste better, encouraging you to eat and buy more. This is the third and last type of taste receptor that leaves you wanting more.

Nature has invested a great deal more variety and creativity in making sure that you detect and reject certain foods. One example is a second type of taste receptor that detects sodium salts when they are overly abundant. These receptors also detect other types of salts in your food, such as lithium salts, which can be toxic. When activated, these receptors signal the brain to create an unpleasant taste sensation and trigger the urge to spit out whatever is in your mouth.

Another type of receptor, the sour-taste receptor, detects acids in your food. If microbes have beaten you to your meal, causing the food to ferment or spoil, the levels of acid in the food will be raised. When you take a bite, your taste receptors sense the acid and send a message to your brain, so that you experience a taste that is unexpectedly sour. Newborn babies naturally reject sour foods. As we get older, we may learn to appreciate a touch of sourness in certain foods, such as lemonade or sweet-and-sour sauce. Yet even for adults, if a food tastes uncharacteristically sour, this is a signal that it has gone bad and should not be eaten.

Finally, there is bitter taste. Your mouth and tongue are home to about thirty different varieties of bitter-taste receptors, compared to one or two varieties *each* for the other basic tastes. This panoply of bitter-taste receptors protects you from eating a range of noxious and poisonous substances. Young children universally reject bitter-tasting foods. As we age and become exposed to culturally accepted bitter foods, such as beer, coffee, and some vegetables, we may learn to appreciate a degree of bitterness. But a substance's intensely bitter taste triggers the same reaction in human children and adults, not to mention dogs, rats, and a host of other animals. Each will make a facial expression indicating disgust and stick their tongue out of their mouth, as if to eject the offending substance.

A surprising fact about this type of reaction, and indeed everything about the sense of taste, is that it is arbitrary. Bitter and sweet tastes are not inherent properties of food. For instance, certain compounds called beta-glucopyranosides have a powerfully bitter taste to humans, whereas mice do not taste them at all. Yet when scientists created genetically altered mice whose tongues grew the human version of a bitter-taste receptor, those mice could taste the compound as bitter and rejected it. Likewise, the sweet-taste receptors of mice do not recognize aspartame, the artificial sweetener found in most diet sodas. To mice, aspartame has no taste, and for them diet sodas have far less appeal. But when scientists created mice with human versions of the sweet-taste receptors, the mice experienced a sweet taste from aspartame and feasted on it. By swapping part of the mouse's sweet-taste receptor with one of its bitter-taste receptors, scientists even created a line of mice that devoured food that normal mice would reject as too bitter. Rather than having an aversion to the food, these mice sought it out.

Ultimately, there is nothing about aspartame that is inherently sweet. There is nothing about *sugar* that is inherently sweet. Sweetness is a kind of label, a category that our tongues and brains use to make swift, safe decisions about what we should ingest. The idiosyncrasies of receptors in your mouth create a small yet important set of taste categories. But, of course, it is your brain that must translate those categories into what you experience as taste. What happens to taste information when it finishes its journey from your mouth to your brain?

TASTE IN THE BRAIN

Sadly, much of what we have learned about the brain over the course of history has come from observing how people or creatures have been affected by brain damage. Nothing reveals the importance of an area of the brain better than its destruction. To see for yourself how one particular brain area contributes to the sense of taste, consider the story of a seventy-five-year-old woman I'll call Mary.

Mary was cooking dinner when the right side of her body suddenly felt weak. She slid to the floor, where she lay confused, unable to speak or respond to what those around her were saying. A blood clot lodged in a major artery feeding her brain was blocking the flow of blood. Deprived of oxygen, the neurons in parts of her brain were shutting down and beginning to die.

Mary was taken to the hospital and received a treatment to dissolve the clot so that the artery could bring fresh oxygen to the suffocating cells. Fortunately, the treatment worked, restoring blood flow to her brain. Still, Mary's stroke left lasting brain damage. On a medical scan, her doctors saw an ominous zone of darkness on the left side of her brain, inside a folded part of cortex that includes the primary taste cortex, where the sense of taste is mapped. This darkness indicated that Mary's treatment came too late to save the neurons in this region of her brain. These cells were permanently damaged or dead.

Mary returned home and began to improve, although she still struggled with some everyday tasks. Only then, back at home and eating her usual foods, did she discover that something was terribly wrong with her sense of taste. Everything tasted like dirt. Despite having enjoyed the taste of ham, chicken, potatoes, and vegetables all her life, she could now no longer tolerate them. Wine and coffee too were no longer palatable. These foods still tasted like *something,* just not something she'd call food. Mary found that she had to force herself to eat. She stopped enjoying meals with loved ones and began to feel isolated. Six months after the stroke, she had lost fourteen pounds.

After some experimentation, Mary eventually discovered foods that she could tolerate. She found she liked tomato sauce with pasta or beef. She could drink tea in lieu of coffee. Her ability to enjoy sweet tastes was unaffected by the stroke, so she could still have desserts and chocolate. She adapted to her new palate and was able to stop shedding pounds, but she never regained her previous experience of tastes. Even a year after the stroke, she could not enjoy a dish of chicken and potatoes. These foods, once her favorites, now tasted like sawdust.

Scientists know less about the neural representation of taste than that of vision, touch, and hearing. For historical and anatomical reasons, far fewer research studies have focused on how the brain handles taste. But we do have some exciting clues. We know that taste information travels from the mouth to stations deep in the brain before arriving at a plot of cortex on either side of the brain, where the primary taste cortex resides. In humans, this cortex lies within the insula, a region of cortex buried within a prominent fold on each side of the brain. The brain damage that disrupted Mary's sense of taste was in this region, within the insula in her left hemisphere. And when scientists have electrically stimulated neurons in this area of the insula in either hemisphere, patients reported experiencing nasty, metallic, or acidic taste sensations.

To learn about the layout of the primary taste cortex, it is best to start with what is known from animals. Neuroscientists generally turn to animals to learn about the fine-grained layout of brain areas. For obvious ethical reasons, scientists cannot use invasive methods to study the human brain up close unless there is a medical need for the procedure. Brain scanning technologies like functional MRI allow us to observe brain activity in humans without harming anyone. But these methods have poor spatial resolution, which means they detect a blurred signature of activity from lots of neurons at once. Compared to the direct methods neuroscientists can use to observe brain activity in animals, noninvasive scanning technologies like functional MRI are a bit like gazing at the brain without one's glasses on.

When scientists have directly studied taste in animals, they have reported intriguing results. One landmark study of mice offers a particularly clear glimpse of a taste map in the creature's primary taste cortex. (See Figure 22.) The scientists used a precise molecular technique to watch neurons in the living brain fire while they fed the anesthetized mice sweet-, umami-, salty-, sour-, and bitter-tasting chemical compounds. They discovered a map with zones, or neighborhoods, representing the different tastes. The zones formed a sort of elongated diamond shape, with bitter and sweet zones farthest apart, and salty and umami zones in between them. The scientists could not locate the sour zone and suspected that it was beyond the patch of tissue that they could view in their experiment.

A second study showed how important the districts of this map were for the mouse's experience of taste. Here, researchers used a different technique to make neurons in specific taste neighborhoods of the map more active by shining a laser directly on those regions of the exposed brain. Remarkably, they were able to do this while the mouse was awake and engaging in its normal behavior. While each

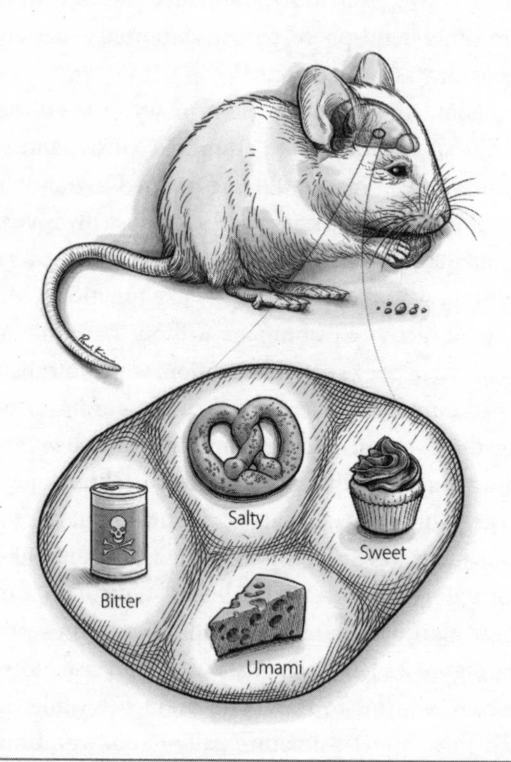

FIGURE 22. An illustration of the taste zones observed in the primary taste cortex of the mouse. ***Paul Kim***

mouse licked water from a spout, the scientists used the laser to trigger activity in the bitter-taste zone of its primary taste cortex. Although the mice were drinking only water, they reacted as though the water tasted powerfully bitter. They stuck out their tongue, gagged, and tried to wipe the offending taste out of their mouth with their paws. In contrast, when the scientists used the laser to stimulate each animal's sweet-taste zone, the mice licked fiercely at the spout, as if feeding on syrup rather than water.

Despite these exciting findings, scientists are still wrestling with understanding the nature and layout of taste maps in mice and other rodents. Some studies have found that the taste districts of the map overlap and that the region is organized around taste pleasantness more than individual types of taste, like sweet or sour. Other work indicates that many of the neurons in the taste map actually respond to other properties of food, such as its aroma, texture, and temperature.

For all that remains unknown about the taste maps of animals, even less is known with certainty about the human taste map. Experiments attempting to study the organization of the human taste map by using functional MRI have yielded conflicting and ambiguous results. Some studies found that the human primary taste cortex features zones for tastes but that these zones overlap quite a bit. Other studies carried out with fancier techniques suggest something else entirely: the presence of taste representation in the absence of any map at all.

This might seem surprising, given how assiduously the brain represents many types of information using brain maps: for instance, the by-now-familiar example of how spatial information like distance and location within a brain map corresponds to information about important events unfolding around us. But although the brain is chock full of brain maps, it can represent information in another way: with a distributed code.

Representing information with a distributed code is entirely different from representing it with brain maps. In a brain map, neighboring neurons represent neighboring regions in space, frequency, time, and so forth. And brain maps essentially represent information using location, or *where* in that brain area the neurons are most active. In contrast, there is no consistent relationship between neighboring neurons in brain areas that use distributed codes. These areas represent information through the distributed pattern of activity across that entire region of the brain, rather than through the location of activity within it. This pattern of activity is a kind of code.

What do I mean by a code, and how does it differ from a map? Let's imagine I've invited you to a party and need to send you directions to the venue. This is a problem of representation: I have information (your route to the venue) that I need to represent on paper or in an electronic message. If I do this well, you will understand the directions based on the message and make it to my party. I could send those directions in one of two ways. I could draw you a map that shows the route from your house to the venue. *Or* I could write out a verbal description of the streets you should take, where to turn, and so on.

If I opt for the second approach, I am using a code to represent and transmit the information. Language, whether spoken or written, is the quintessential example of a code. Consider how we form written words out of letters. Many written languages are based on an alphabet, or a small set of letter symbols. To make meaning out of such letters, I must combine them into words. It is only by combining these letters that I can create unique letter patterns, or words, that represent and mean something to you and to me.

In brain areas that use a distributed code, the activity of individual neurons plays the role that individual letters do in written languages with an alphabet. A single neuron can fire quickly in response to many different things, just as a single letter can be used to form many different words. For distributed codes, what matters is the specific set of neurons that are firing like mad at that moment. Information is contained in the *pattern* of activity across many neurons rather than in the activity of any single neuron.

Brain maps are a useful way for brains to represent information because they make representation and information processing efficient and compact. But what is the benefit to representing information with a code? In a word, that benefit is flexibility. Any given map has fixed dimensions and boundaries. In the case of brain maps, those could be surfaces of the body or regions of visual space. Every part of the map is assigned to representing something specific, which leaves no room in the map for representing *new* things, such as touch on an entirely new body part or vision from an eye newly installed in the back of your head. Likewise, if the venue for my party were moved to a new city that was beyond the boundaries of the original map I made for you, the old map would be of no help. I would have to draw an entirely new map to get you to my party. In short, maps do not handle *new* things well.

Codes do not have this problem. I can use the existing alphabet to create new words or combinations of words to convey new meaning. Venue moved? No problem! I can create new directions for you using the same set of letters, only in different combinations. This kind of flexibility is important for the brain, particularly for phenomena in which *newness* is common. Your brain can simply create new patterns of neural activity to represent new tastes, objects, or locales as you encounter them.

Although maps and codes are opposites in many ways, they are not in opposition. They operate jointly to support nearly everything you do. For example, you use frequency-based maps like A1 *and* distributed codes to convert sound-pressure waves detected at the ear into voices that you recognize (*That's my mom!*) and words you comprehend (*She's calling me to come home!*). Maps and distributed codes generally exist in separate areas of the brain that work together by sending signals back and forth. But maps and codes can also be combined in certain parts of the brain, particularly those that contain maps with zones, like the taste zones in the primary taste cortex. For example, neurons within a sweet-taste zone might use a distributed code to represent specific features of sweetness. Or neurons nestled in the no-man's land between zones might represent new flavor features using a distributed code. This happy compromise marries the flexibility of codes with the benefits of maps. The human primary taste map may reflect just such a compromise. It will take more exploration on the part of intrepid scientists to know for sure.

SCENT FOR SURVIVAL AND ACTION

As vital as your sense of taste may be, your sense of smell is by far the more impressive and mysterious of your chemical senses. When considered across the animal kingdom, it is nearly impossible to overstate the importance of scent. Sharks, snakes, mosquitoes, vultures, badgers, and hummingbirds are just a few of the creatures that follow their nose to find food. Scent can also signal social status, as it does among termites, which recognize their queen by her scented secretions. It drives reproduction in a stunning variety of ways: the spotted hyena wipes its pungent anal paste on grass to advertise its reproductive status, and the male frillfin goby, a marine fish, launches into an hour-long courtship routine whenever he gets a whiff of a fertile female's ovarian secretions. Scent is also integral to parenting and bonding, enabling many newborn creatures to recognize their mother and drawing newborn mammals to her nipples to nurse. Albatrosses and other seafaring birds even find their way across the vast ocean using their sense of smell. In short, smell is essential for nearly all aspects of animal survival. But how do animals extract the information they need from odors, and what kinds of maps do their brains use to do this?

Olfaction, or the sense of smell, is a remarkable feat of molecular recognition. Consider the example of mouse olfaction. Embedded within the lining of the mouse's nose are about ten million smell receptors made up of about a thousand different types. Any given airborne molecule might bind to more than one type of receptor, and any given type of receptor might bind to more than one kind of molecule. As a result, mice can detect and identify far more than one thousand odors, even though they have only a thousand receptor types.

When airborne molecules bind to the sensory receptors inside a creature's nose, the attached neurons send a signal on to the brain. These signals travel directly to two conspicuous balls of brain stuff called the olfactory bulbs, which jut out in front of the brain in mice, humans, and other animals. The right and left olfactory bulbs each contain a detailed map of odor zones loosely organized around the structures of the molecules they represent, such as how long its carbon chain is or whether it is a carboxylic acid, a phenol, or an aliphatic ester. Although the chemical jargon might mean little to you, this structural information is the key to identifying *what kind of stuff* that molecule came from and therefore how it might be relevant to you. The map in your olfactory bulb takes the first step in this process by specifically representing information about which type of molecule has made its way into your nose.

From the olfactory bulb, information about smell goes to several regions of the brain. Of these, the one scientists have studied most in rodents and humans is a region called the piriform cortex. Experiments have shown that this area plays a crucial role in learning

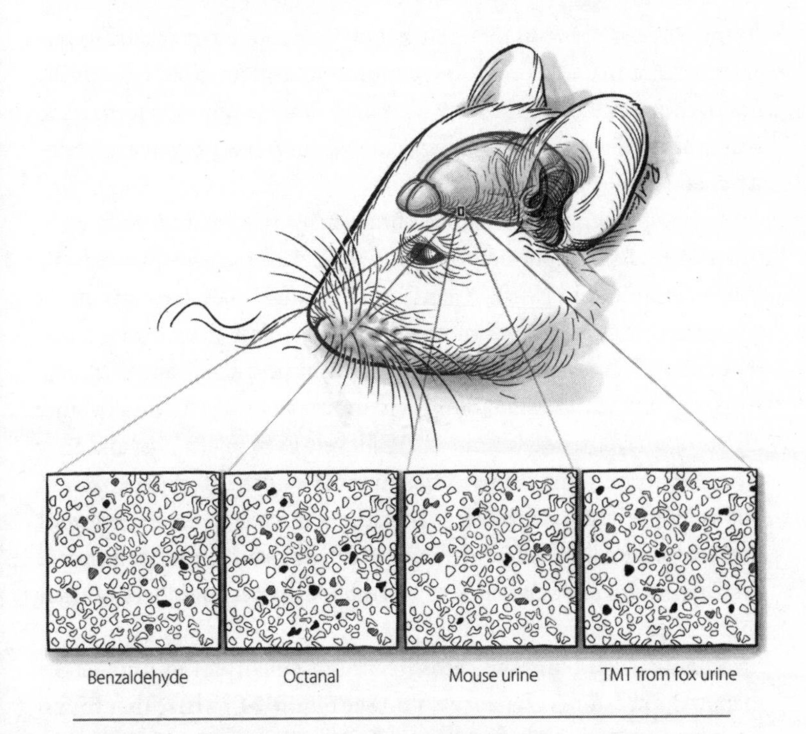

FIGURE 23. An illustration of a distributed code for odor in the piriform cortex. *Paul Kim*

new odors. And as you might imagine, given how poorly maps handle *new stuff*, the piriform cortex represents odor with a distributed code rather than a map. The illustration in Figure 23 shows this code in action, depicting how a set of neurons in the mouse piriform cortex collectively represents different odors through their different patterns of activity. The neurons with high firing rates are indicated with black; those with slightly heightened firing rates are shown in gray.

At this time we know more about the piriform cortex and its distributed code for odor than about other odor-processing regions of the brain. Still, scientists have found a handful of intriguing brain areas beyond the olfactory bulb that are organized into odor zones. One of the challenges in finding such maps is the complexity of olfaction and its truly overwhelming number of detected molecules and detecting receptor types. Above all, the challenge stems from the countless possible ways that odors could be grouped or related to one another in a map. In order for scientists to find an odor map in the brain, they first need to know which map dimensions or categories to search for. This is true for all brain maps, but it has proven a particular challenge for olfaction.

One approach to finding odor brain maps has been to study creatures with less complicated olfaction. For example, the channel catfish, a fish found in North American rivers and lakes, has only about a hundred different types of smell receptors that detect only a few types of molecules, including nucleotides, amino acids, and bile salts. Although nucleotide and amino acid molecules are structurally quite different, they are both found at high levels in living things, and they both mean the same thing to the channel catfish: food. In contrast, bile salts are created by the liver and then released in the feces or urine of other fishes. Like hyenas and their anal paste, catfish use these bile salts as social cues to learn about other members of their species that are nearby.

Scientists studying the catfish sense of smell first examined the map in the creature's olfactory bulb. (See Figure 24.) There they found three separate zones for the three different types of molecules: nucleotides, amino acids, and bile salts. This fits with findings from other creatures: the olfactory bulb map is organized around the structures of odorous molecules. But the catfish's olfactory bulbs send information about smells on to another part of the catfish brain, where the scientists found another odor map. This map contained only two major zones: one for bile salts and one for both nucleotides *and* amino acids.

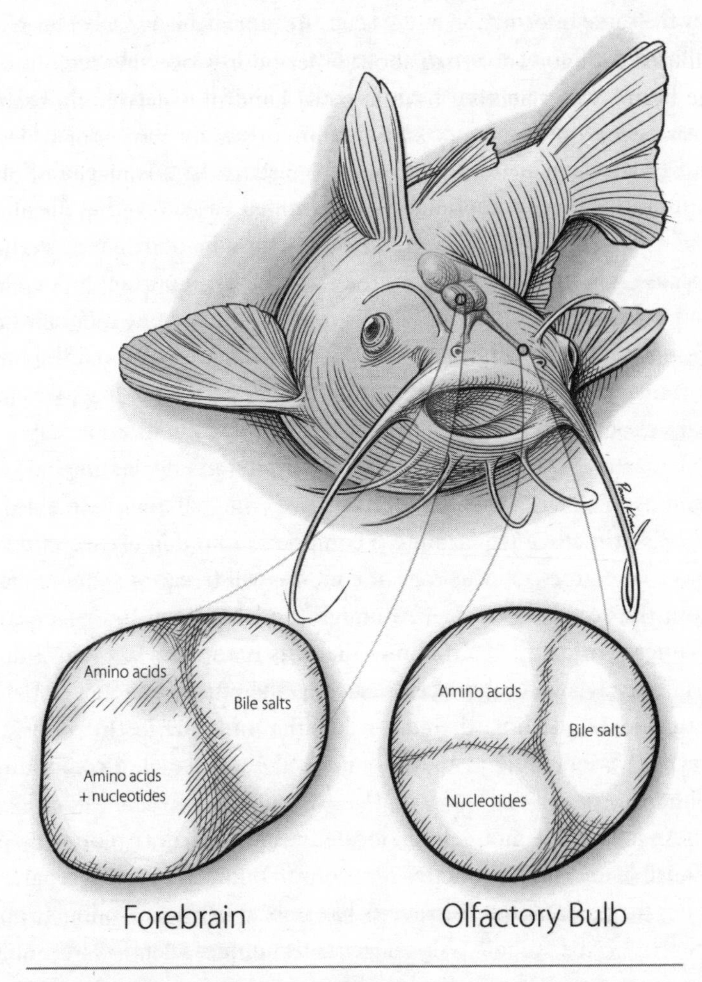

FIGURE 24. Odor maps in the olfactory bulb (right) and forebrain (left) of the channel catfish. *Paul Kim*

The difference between these two odor maps is subtle but important. The map in the olfactory bulb is organized into odor zones based on molecular structure, an objective property of these compounds in the physical world. But the second map is organized into zones based on the significance of the odors to the animal. Whether a catfish smells amino acids or nucleotides, it will launch into the same type of feeding behavior. These two different types of molecules convey the same information and trigger the same behavior, whereas bile salts offer different information and call for different behaviors.

The second odor map of the channel catfish is a map of behavioral relevance. Only compounds that are important for the fish's survival and procreation are awarded zones or districts in this map, and the grouping of these compounds is determined based on what the animal will need to *do* in response to them. This kind of map is powerful because it distills and classifies the stuff that is important to a creature. As you might imagine, such a map would be quite different for a mouse, a lion, a vulture, or you. Just as we can learn about the importance of the pony's nostrils from its distorted touch map, we can learn about the importance of stuff to a catfish from its odor maps.

Odor maps that are important for driving innate, instinctual responses to smells have also been found in mice. If you place a drop of 2,3,5-trimethyl-3-thiazoline, a compound found in fox secretions, into a laboratory mouse's cage, the mouse will freeze or scurry as far from the scent as possible. Although the lab animal has never encountered a fox, its brain knows that this is an odor to avoid. Mice instinctively avoid other scents, such as the musky bouquet of bobcat urine. In addition, there are odors that mice are instinctively attracted to, including peanut oil and 2-phenylethanol, a compound found in rose oil.

An area in the mouse brain located near the piriform cortex is responsible for these instinctive reactions to odors. It contains separate zones for predator odors to avoid, like bobcat urine, and odors to approach, like peanut oil. Using a clever technique, scientists were able to activate neurons in one zone or the other, without exposing the animal to actual odors. When scientists stimulated the predator-scent zone, the mice froze or scuttled away, just as they would if a predator were present. When the scientists stimulated the attractive-scent zone, the mice lingered, as if hoping for a snack. Another study found a nearby zone that responds to the smell of urine from mice of the opposite sex. This region seems to be involved in processing chemical signals that are relevant to mating.

There is still a lot of work to be done in uncovering the organization of the mouse's odor maps, but these studies suggest at least three zones: one for predators, one for food, and one for mates. Each zone has clear behavioral relevance and appears to bridge the connection between which compounds the mouse detects and which action it makes in response.

These studies offer a glimpse of odor maps and the behavioral relevance of scent to catfish and mice. But what about humans? In order for scientists to find odor maps in humans, they need to know what kind of map zones or dimensions to look for. In animals, the layout of a creature's odor maps is determined by *how* it uses odor information and, specifically, how that information leads to *action* related to reproduction or survival. Which raises this question: how *do* humans use smell to survive, if they do so at all?

My mother, Sally, lived her entire life without ever once smelling a thing. So far as anyone can tell, she was born without the capacity to detect odors of any kind. She wasn't even sure that she could imagine what smelling was *like*. After all, how can people go about imagining a sense that they have never actually experienced?

Sally would talk about her missing sense of smell like it was a curiosity or a philosophical puzzle—but never a handicap. Its absence simply did not seem to affect her very much. Some of her acquaintances weren't even aware that she lacked one of her five senses. By contrast, imagine not knowing that your friend couldn't see or hear. Sally faced a few dangers because of her lack of smell; she could not smell a gas leak, could not smell smoke (although she could feel it with her eyes and nostrils when the smoke was concentrated), and could not tell that milk had spoiled until it poured from the carton in curdled chunks. But for the most part, Sally's lack of smell made her seem stoic. She was unfazed by a reeking outhouse, just as she was unmoved by the smell of freshly baked bread wafting from a nearby bakery. In some ways, her missing sense seemed less a disability than a superpower. But how could that possibly be?

It's important to distinguish between the experiences of people born without one of the five senses and those who lose it later in life. People who lose their sense of smell later in life, after they have come to associate foods, places, and experiences with certain scents, tend to find this loss extremely distressing. Yet aside from the substantial emotional impact, these individuals do not usually find themselves incapacitated. The loss does not leave them unable to navigate from place to place, fulfill the requirements of their jobs, or otherwise function independently. The same cannot be said for people who abruptly lose the sense of sight or hearing. The logical conclusion would seem to be that smell just isn't that important for humans.

Until relatively recently, the scientific community endorsed the same conclusion. It was widely believed that humans have a paltry sense of smell compared to that of other creatures in the animal kingdom. Some proposed that the evolution of the human brain has been marked by a neural divestment in the sense of smell. To support this claim, they pointed to the olfactory bulbs. One thing that struck early anatomists is how small human olfactory bulbs are, relative to our large brain. As you can see in Figure 25, the olfactory bulbs of the mouse make up a much larger chunk of its total brain size. Scientists concluded that, as the brains of modern humans evolved, the growth of their olfactory bulbs was stunted and our ancestors' ability to smell declined as a result. Some claimed that this happened because humans rely on reason rather than reflexive reactions to odors. Others suggested that our ancestors evolved to have brains that invested in vision at the expense of smell, making us super seers and pathetic smellers.

Despite these long-held notions, recent decades have brought a surprising revelation: humans are far better smellers than we ever imagined. Head-to-head tests of perception across species have shown that the human sense of smell is on par with that of the mouse. Although mice beat out humans at detecting certain odors, humans best mice on others. We also tie with dogs and rabbits in some odor matchups, and even surpass them in others. Moreover, studies that actually count neurons in the brains of various species suggest that mice, humans, and many other mammalian species all have about the same total number of neurons in their olfactory bulbs, regardless of the absolute or relative size of these structures.

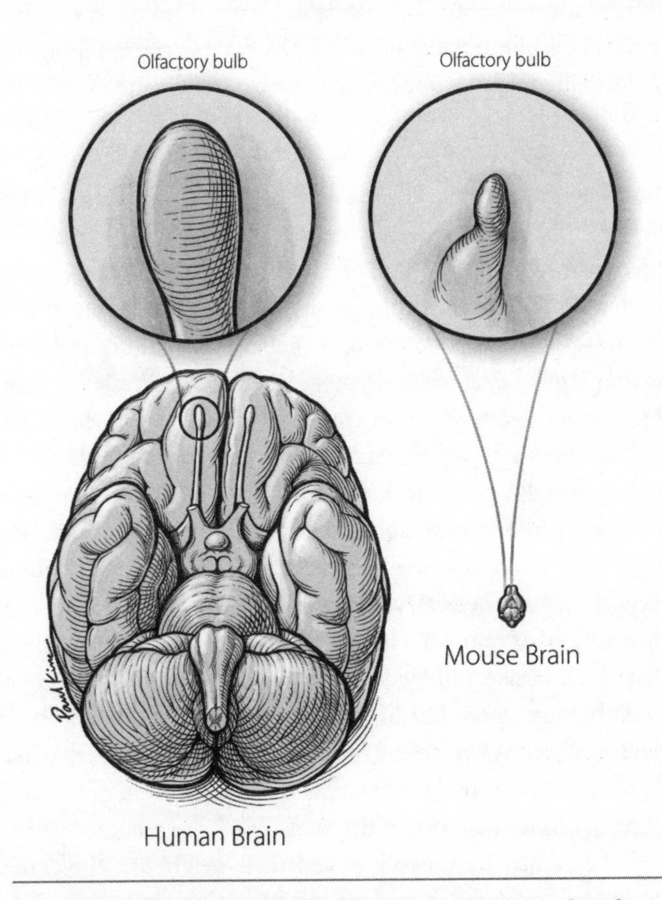

FIGURE 25. A comparison of olfactory bulb size (above) and total brain size (below) for humans and mice. ***Paul Kim***

Humans, like other creatures, are privy to a wealth of odors that inform us about stuff that is nearby. We can detect 1-octen-3-ol, a compound in mushrooms that yields their familiar scent. We can smell 2-isobutyl-3-methoxypyrazine from bell peppers, vanillin from vanilla, and eugenol from cloves and from wine that has soaked in oak barrels. We can enjoy the smell of geosmin, a compound produced by microorganisms in dirt, which gives us the earthy smell of soil after a summer rain. We can suffer the smell of compounds like trimethylamine from deteriorating fish carcasses and tetrahydropyridine from rotting meats and vegetables. The world of scent is filled with rich detail about the objects and physical processes taking place around us. But does this information drive our daily life-or-death decisions? Based on how well people manage after losing their sense of smell, the answer would seem to be no.

But there is a missing variable in this equation. Nearly all of our studies of human olfaction, like the vast majority of psychology and neuroscience research on humans, has been carried out on people from a handful of Western cultures with their own idiosyncratic way of life. In fact, there is good reason to believe that culture and way of life matter a great deal when it comes to olfaction. Westerners don't tend to use the nose for finding food, but this is not true of people from all cultures and was almost certainly not true of our distant human ancestors. People from hunter-gatherer societies rely more on odor to glean information about their environment and have clear categories for describing smells. Among the Jahai, a group of hunter-gatherers living on the Malay Peninsula, the word *cŋes* is used to describe a variety of odors, including those of gasoline, bat droppings and bat caves, smoke, millipedes, wild ginger root, leaf of gingerwort, and wild mango wood. Mushrooms, deadwood, stale food, fur, and feathers are *pʔus*. When scientists asked members of the Jahai and a group of Americans to identify scents familiar to Westerners, the Jahai blew the Americans out of the water. Likewise a hunter-gatherer society, the Semaq Beri, easily bested their neighbors, the agricultural Semelai peoples, at odor naming, even though the two groups spoke related languages. In other words, smell is discussed, used, and conceived of differently in different cultures, which are shaped by different ways of life. If an individual, like my mother, without a sense of smell were born into a hunter-gatherer culture, she might have been severely handicapped. And just as the behavioral relevance of smell can differ between cultures, so too might the layout of odor maps in the brain.

The long debate over our human capacity to smell presupposes that olfaction must be a means of gathering information that we can consciously perceive and report. But the dirty little secret about olfaction—in Western cultures and elsewhere—is that it works much of its magic without our ever knowing it. In fact, the value of smell for informing us about the stuff around us may pale in comparison to its value as a delivery service for secret messages.

You deliver such messages all the time, although you don't realize it. Humans are home to myriad glands in the skin that harbor bacteria. You don't need to be sweaty from exercise to give off body odors; small amounts of these secretions are oozing out of your glands pretty much continually. People may not notice or comment on your body odor, but that does not mean that you don't have one or that they are not detecting it, at least subconsciously. A growing number of studies show that the body odors we give off contain tons of information about ourselves—about our gender, age, physical health, emotional state, and fertility. Although scientists don't yet know exactly which compounds in the odors convey these specific details, they can tell that these cues exist because of their effect on the people who smell them.

For example, when others smell the odors that you give off when you are fearful or anxious, they can quite literally smell your fear, even if they do not consciously realize it. And smelling your fearful sweat will make *them* more likely to detect threats in their environment. It can even influence their performance in high-stress situations, making people more likely to bungle a procedure or exam. Findings like these suggest that our bodies are constantly sending messages to one another via scent—messages that people do not consciously send nor consciously receive, even though they alert us to potential dangers (illness or threat) or opportunities (a fertile potential mate) and influence our behavior.

Communicating fear is just one way that humans influence one another through scent. Body odor also communicates identity; newborn infants recognize their mother's scent, and family members learn to recognize their infant's scent soon after birth. Body odor also causes young women living in close quarters to synchronize their menstrual cycles. The scent of human tears reduces human males' sexual attraction to females. And people unconsciously tend to sniff their fingers after shaking hands with a stranger, although what they are sniffing for, and why, remains an intriguing mystery.

Scientists are only just beginning to discover how we actually use smell to guide our interactions and behavior. They know even less about the role that culture and lifestyle play in this process. And so perhaps it should come as no surprise that scientists have yet to discover human brain maps for odor beyond those in our olfactory bulbs. Of course, we may have failed to find them because there are simply none to find. But until scientists have a better understanding of how humans actually *use* odor, universally and in specific different cultures, they will have little chance of unearthing maps that make it possible.

6

On the Move: Brain Maps for Action

THE VERY FIRST DISCOVERY OF A MAP in the human brain arose from the strange and horrible experiences of certain patients in nineteenth-century London. History remembers them as a list: partial names or initials, ages, and body parts in outright rebellion. For one fifty-two-year-old man, known to the annals of medicine only as "C.," the convulsions started in his left big toe. From there, they spread to the inner side of his foot, up his leg, and eventually, after he had lost consciousness, to his arms and face.

For nine-year-old Elizabeth F., the seizures would begin in her right eye when she was talking or singing. The eye would twitch, and then her mouth would open and her face would draw to the right. Her hand would clutch her head, and her leg would kick. She could not speak during the seizures, nor for several minutes or sometimes hours afterward.

There was James R., age thirty-nine, whose seizures began in the right hand; he also reported a terrible pain in his head. He died twelve days after his admission to the hospital. His autopsy revealed a growth on the left side of his frontal lobe. Once excised, the tumor measured a cubic inch in size; the doctor described it as bluish on the outside but dead and gray within.

And there was a twenty-two-year-old man, described only as "fairly nourished," whose seizures were triggered by coughing fits brought on by tuberculosis. One morning, after finishing his breakfast and suffering a fit of coughing, the man's left thumb began to move back and forth as if of its own volition. After several seconds of this, he felt what he described as a painful numbness spread over his entire body, and he lost consciousness. It was the first of many seizures, and they always began in his left thumb and then spread up his arm and throughout his body. About six weeks after entering the hospital, the poor man died of tuberculosis. An autopsy found "a tubercle the size of a hazel-nut" in his right frontal cortex. The doctor pried the tumor quite easily out of the brain tissue and sliced it open, finding the center "slightly cheesy."

Aside from the miserable tales of their maladies, these patients had loved ones and meaningful lives. But from the perspective of medicine and science, they presented opportunities to learn about human physiology and disease. Those mentioned here were but a handful of the hundreds of patients who were examined and described by John Hughlings Jackson, a renowned British neurologist of the nineteenth century. Jackson was well known to the London medical scene as a dry lecturer but a keenly observant clinician. He was also famous for his absentmindedness. It is said that at a dinner party, he took his handkerchief from his pocket to blow his nose, whereupon a large chunk of brain tissue tumbled out onto the table.

At the time that Jackson was studying these patients, there was not much he could do to improve their lives or treat their maladies. But he could learn from their stories, observe their symptoms, and describe the common themes he found among them. And what he found was a *path* that seizures consistently traveled as they spread through the body. He wrote, "When fits begins [*sic*] in the hand, the [seizure] spreads up the arm and down the leg; when a fit begins in the foot it spreads up the leg and down the arm."

The type of seizures Jackson described came to be known as Jacksonian seizures. And the pattern of convulsions that he observed became known as the Jacksonian march. The name implied an analogy—convulsions are like an army marching along a certain path. They might start at different places along that path for different patients (for example, C.'s left toe or Elizabeth F.'s right eye), and they might march in one direction or the other—or even in two directions at once. But because of these established pathways, they will always reach certain landmarks in the same order. A seizure starting in the right leg must first spread to the right hand before it can reach the right side of the face. Convulsions don't jump from the toe to the cheek. Likewise, they might start in both the left fingers and left cheek, but they would not start in both the left and right fingers.

Jackson observed these commonalities in how and where seizures struck his patients and, from them, he concluded that there is a special layout within the human brain related to movements of different body parts. He envisioned a path in the brain that could trigger movements in other parts of the body. This path was a map of the body laid out in the following order: toe, leg, torso, arm, hand, fingers, face, and head. Since seizures tended to spread along the same side of the body and since obvious damage to one side of the brain could cause seizures on the opposite side of the body, he deduced that the map was split into two parts. One half of the map, in the left hemisphere of the brain, primarily controls movement on the right side of the body, and the other half, in the right hemisphere, is in charge of movement on the left side of the body. Together, these swaths of the brain make up a complete map of the human body.

In a healthy brain, this map helps us move our body according to our will. But if a tumor or abscess penetrates that map, creating an instability somewhere within it, that instability can generate movement—convulsions—in the body part corresponding to the affected part of the map. And if that activity spreads outward from that point in the body-movement brain map, the convulsions would spread across the patient's body accordingly. For example, instability from a tumor where the right foot was controlled would spread from the right foot to the right leg and upward to the right hand and face. In more extensive seizures, the unstable activity would also spread to the map on the opposite side of the brain, so that the seizures would overtake both sides of the body.

Jackson's keen clinical observations offered other clues to the layout of this movement map in the brain. He observed that, for most of his patients, the seizures began in the hand. The second most common starting place was the face or tongue. It was far less common for Jackson to encounter patients whose seizures began in the foot. Likewise, when a patient's seizures started in the hand, they tended to start in the thumb or index finger and not in the middle, ring, or pinky finger. And for those patients whose seizures *did* start in the foot, Jackson observed that the fits always began in the big toe. Overall, he noted that seizures are most likely to start in those parts of the body that have the most mobility and that we use the most in our daily physical activities.

Jackson's observation tells us something about size. A tumor or abscess that randomly develops somewhere in a person's movement map is more likely to develop in its largest regions rather than the smallest, just as a meteor falling to earth is more likely to strike India than Luxembourg simply because India is larger. In noting that seizures tend to start in the thumb, index finger, face, or tongue, Jackson uncovered the first evidence that the corresponding regions of the body-movement map are especially large. In other words, the movement map is, like other brain maps, warped by magnification.

When Jackson's patients died in his care, he could sometimes perform an autopsy, as he did for James R. and the man with tuberculosis. Often Jackson found abscesses or tumors in a particular fold of the frontal lobe, on the side of the brain opposite to the body part where the patient's seizures began. This region of the frontal cortex would come to be known as the primary motor cortex, or M1. Around the same time, experiments with animals were yielding evidence for remarkably similar movement maps in the frontal lobes of dogs, monkeys, rabbits and other animals. It seemed that Jackson's observations and deductions were on the right track.

Around 1872, Jackson applied what he had learned from James R.'s brain tumor to predict the location of a tumor in a new patient: a woman with frequent seizures of the right hand and arm. When she died after a severe seizure, Jackson's autopsy confirmed that his guess had been right: her tumor was in the hand region of the left motor cortex.

Jackson's method of using seizure patterns to pinpoint damage in the brain had a profound and immediate impact on medicine. Before that time, surgeons generally did not operate on the brain, in part because they had no idea which part of the brain was causing a particular patient's symptoms. As the respected Scottish surgeon William MacEwen described it, "The brain was a dark continent, in which they could descry neither path nor guide capable of leading them to a particular diseased area, and did they attempt to reach it, it could only be by groping in the dark." Jackson's observations of his patients' seizures formed the crucial first step in sweeping away this darkness. Soon after he told the world about the human M1 map, surgeons began using a patient's symptoms and their knowledge of this map to guide successful brain surgeries.

William MacEwen himself was among the first to take that next step when, in 1879, he used Jackson's descriptions of the layout of the M1 map to save a boy's life. MacEwen's patient had taken a big fall, bruising his head and face, and six days later he began to have seizures. Each one started with twitching on the left side of the face, before the convulsions spread to the left arm and then the left leg. MacEwen recognized the boy's seizures as a Jacksonian march. Because the seizures began on the left side of the boy's face, McEwen deduced that the inciting injury in the brain must be located within the face region of the boy's movement map in his right hemisphere.

Based on autopsies like those Jackson carried out, McEwen knew where to find this face region: in the mid-to-lower portion of the right motor cortex. He opened up the skull there and found that the boy had fractured his skull when he fell, injuring the brain tissue underneath. McEwen drained two ounces of fluid and coagulated blood from the wound, replaced the skull, and sutured the scalp. The boy fully recovered, seizure- and symptom-free. This remarkable success was just the first of many effective brain surgeries in the 1870s and '80s. At a time when horseless coaches and light bulbs were making their first appearance on Planet Earth, Dr. MacEwen was successfully carving tumors and infected abscesses out of the human brain. Before the invention of CT and MRI machines, he used knowledge of brain maps to guide his scalpel and save his patients' lives.

Nearly half a century after MacEwen's breakthrough, another pioneering neurosurgeon, Wilder Penfield, would unearth new details about the M1 map and how movement is represented in the human brain. His observations would raise as many questions as they answered. Penfield made these observations while he electrically stimulated regions of the brain in patients who were awake and alert. The goal of the procedure was to guide his surgeries and treat patients suffering from seizures. But in the process, he had a front-row seat to watch human brain maps in action.

Penfield's exploration of the M1 map mirrored his now-familiar exploration of the S1 touch map. The S1 touch map lies right behind a prominent crevice called the central sulcus, which runs across the top of the brain. If you run your finger across your head, from the top of one ear to the top of the other ear, you are roughly outlining the path that the central sulcus cuts across your brain. The M1 movement map lies just in front of this line. In fact, the M1 and S1 maps lie alongside each other, forming the two banks of the central sulcus. When Penfield operated on patients who suffered from Jacksonian seizures, he typically explored both sides of the central sulcus, stimulating different regions of both the touch and the motor cortex and recording what his patients felt or did in response.

In 1937, Penfield and his colleague summarized their observations about the layout of the motor cortex. Although his summary was based on more than a hundred patients, it is easier to consider his findings in the context of a single person. Patient F. W., who we will call Fred, was an "intelligent and cooperative" boy who suffered from seizures that started in his right hand. Penfield's surgical team removed a portion of the boy's skull on the left side and began to probe both the M1 movement map and the neighboring S1 touch map. When stimulating a site caused a clear sensation or action in Fred, Penfield labeled it with a small square of paper laid directly on the surface of the boy's brain. These paper labels allowed Penfield to keep track of the different sites throughout the procedure.

The photograph in Figure 26 shows the labels Penfield placed on the surface of Fred's brain during his surgery. The dark, serpentine dividing line is a blood vessel resting atop the central sulcus. To the left of the central sulcus is the boy's motor cortex; to the right, his touch cortex.

Penfield found that the layout of Fred's M1 movement map reflected what he had found in other patients. Near the base of the map are sites that control movements of the tongue, mouth, throat, and jaw. When Penfield stimulated this area of Fred's brain—a site labeled C, which is not visible in the photograph—the boy's lips moved and his throat issued a soft involuntary noise. Other patients stimulated in this general area might smack their lips, yell, suck, or swallow. Slightly farther up in the map is a region that causes face movements. When Penfield stimulated a site labeled B (also not visible in the photograph), Fred involuntarily closed his eyes. Other patients reacted to stimulation of this area with eye twitching or movements of the nose or eyebrow. Above the areas related to the mouth and face lies the region of the map that triggers finger and hand movements. When Penfield stimulated site 18, Fred's arm and hand twitched. Stimulation of this area in other patients could cause their fingers to flex, extend, or twitch. Finally, at the top of the brain, Penfield found sites that triggered movements of the lower extremities. Stimulation of the site labeled G caused Fred's right knee to flex.

Overall, Penfield's observations of the layout of the human M1 movement map supported and extended what Jackson had deduced more than fifty years prior. But Penfield also encountered plenty of

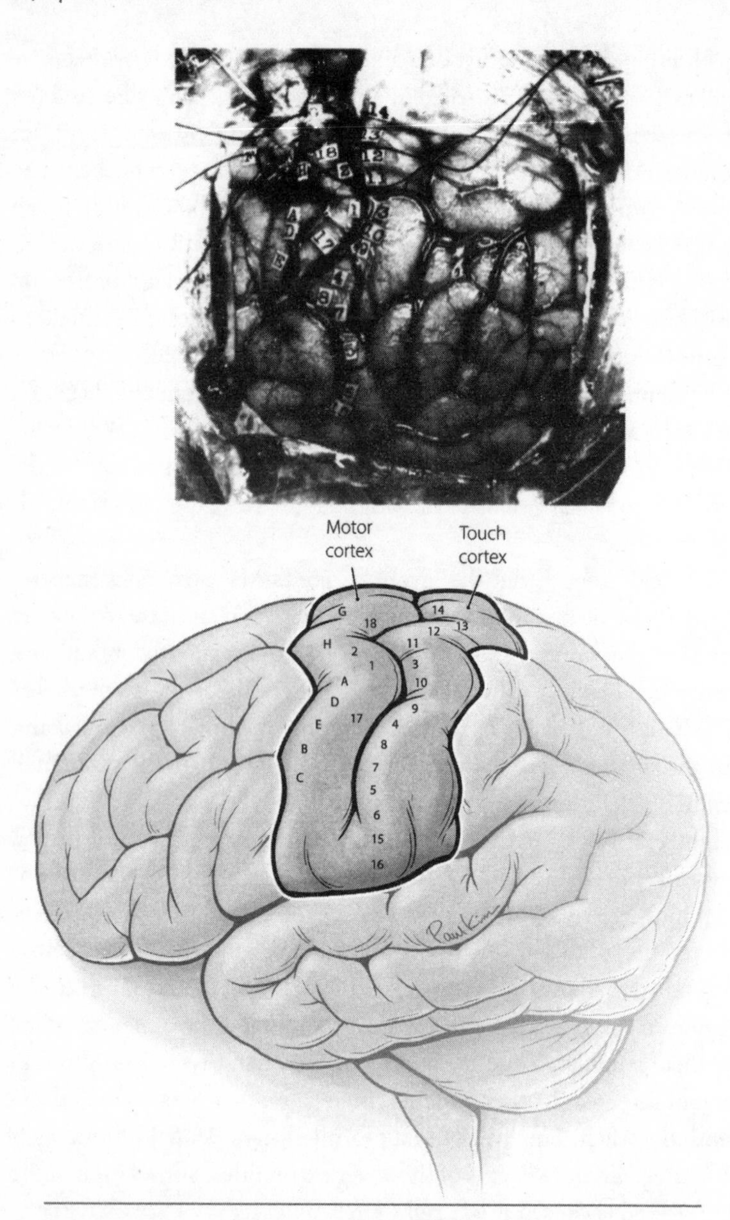

FIGURE 26. **A photograph (above) and illustration (below) of the sites Wilder Penfield probed in Fred's brain.** *Photograph from* Brain: A Journal of Neurology, *vol. 60, no. 4. Copyright © 1937 by Oxford University Press. Illustration by Paul Kim.*

contradictions. For example, stimulation of sites D, 1, and 18 in Fred's brain all triggered movements of his hand. Yet intermingled with these sites were site A, which triggered an arm movement, and site 2, which elicited a shoulder twitch. Although the overall rough layout of the movement map matched Jackson's predictions, examination at a finer scale showed that M1 lacks a smooth one-to-one mapping between body part and cortex. In a word, the movement map was surprisingly *messy.*

In the process of probing Fred's motor cortex, Penfield found a small patch of abnormal hardened tissue that triggered the boy's seizures. He cut out this damaged tissue and sealed the boy's brain back within the dark vault of his skull. Penfield's written report tells us nothing of what happened to Fred after his surgery. Whatever became of his life and his illness, they are lost to us. What endures is the annotated landscape of his brain and the clues it ceded about the movement map contained within.

Penfield's work left many important questions about the M1 brain map unanswered. Stimulation with his electrode triggered a host of different movements in his patients, ranging from twitches and jerks to vocalizations and gestures that required the coordinated action of many muscles at once. The sheer variety of these movements, along with the messiness and apparent contradictions within the map, raised important questions: *What is actually represented in the M1 map?* Is the M1 movement map a representation of the muscles in many individual body parts, all ready and waiting to twitch, or does it represent entire actions according to some different dimension? It would take scientists another sixty years to answer these questions and reveal the surprising nature of the M1 map.

MAPPING ACTION SPACE

The young scientists who revolutionized our understanding of the motor cortex originally set out not to do something new, but rather to do things differently. It was the dawn of the twenty-first century. Like many neuroscientists before them, they guided their tiny electrodes to the region of the monkey's M1 map that triggers movements in the hand on the opposite side of the body. Thanks to more than a century of research on the motor cortex, they already knew what would happen if they sent a brief electrical jolt down the electrode and into the animal's brain. By now it was firmly established that zapping M1 for less than a twentieth of a second would cause a creature's corresponding body part to twitch. But these young scientists were driven by different questions: What would happen if they prolonged the zapping of M1? What if they stimulated the tissue for a full half-second? What would the animal do?

There were several good reasons for the scientists to raise this issue. One was that it takes time for bodies to carry out movements. One-twentieth of a second is long enough for an eye to blink or a muscle to jerk, but it is not long enough to make a full intentional movement, such as reaching out to grasp a nearby object. Perhaps, by stimulating M1 for so short a time, scientists and surgeons had been triggering the start of a larger movement, only to cut it short an instant later when the stimulation ended. Penfield himself recognized this possibility as early as 1951, when he wrote, "The stimulating electrode has frequently been removed at the first evidence of response, and thus the opportunity of producing more of the elaborate synergic responses may have been missed. If this surmise is correct, it would be proper to consider many of the motor effects as fractions of an uncompleted complex." In fact, electrical stimulation of the cortex occasionally offered brief glimpses of rich, coordinated actions in both animals and humans. Were they flukes, or were coordinated actions, like reaching and grasping, somehow etched into the M1 movement map?

The team of young scientists at Princeton University—Michael Graziano, Charlotte Taylor, and Tirin Moore—set out to answer these questions. They placed an electrode in the hand region of the monkey's motor cortex and stimulated each site for a full half-second—ten times longer than most prior experiments had stimulated. The scientists were astonished by what they saw. When they stimulated the monkey's right motor cortex, its left thumb and index finger pressed together as if plucking something from the air, while its arm, elbow, and shoulder rotated to bring the left hand to the face; the animal's mouth opened as if to accept the unseen morsel it had plucked. This hand-to-mouth movement was no twitch. It involved the smooth coordination of many body parts and muscle groups at one time.

Michael Graziano later described the team's surprise and excitement when they watched this remarkable action. "The first time we found a hand-to-mouth cortical site and realized that we could trigger the movement on demand with the press of a button, we were so surprised that we ran out of the experiment room and tried to find someone else in the building, anyone, to watch and make sure that we were not going crazy."

Stimulating a different spot in M1 caused the monkey to act as if something foreign had touched it on the right cheek. The animal closed its right eye, turned its head to the left, shrugged its right shoulder, and lifted its right hand as if to protect its face. Through it all, the monkey seemed unfazed, and kept using its other hand to feed on fruit snacks. Stimulation to other parts of the map caused the monkey to reach for nonexistent objects, to chew, to climb, or to leap. Each action was coordinated and appeared purposeful, even though it was triggered at the scientists' whim, with the press of a button.

As the scientists explored the whole M1 map, they noticed that it is divided into general zones, which are depicted in Figure 27. Stimulation of one zone at the bottom of the map, roughly corresponding to what Jackson or Penfield would have called the mouth representation, caused animals to chew or lick, as if they were eating. Stimulation of another zone, located around the supposed leg region of the map, caused the monkeys to begin jumping or climbing. Another zone, which partially overlapped with the classic hand region, triggered defensive movements. When the scientists stimulated the brain within this zone, the monkey might shield its face, hide its arm behind its back, or flinch as if in response to an unseen predator.

The researchers also found zones devoted to reaching within the sizable classic hand and arm areas of the motor cortex. When they stimulated sites within this zone, the monkey would always reach with its opposite hand, but the exact location of the stimulation within this zone determined where in space the monkey would reach. For example, if they stimulated one part of the reach map in the left motor cortex, the animal's right hand would always reach directly out in front of its body. Stimulation of another site would make the animal reach elsewhere, such as up and to the left.

The truly remarkable thing about this miniature reaching map nestled within the monkey's larger movement map is that it did not represent specific movements so much as their endpoints. For example, the exact movement elicited by stimulating a reach-out-in-front-of-the-body site depended on where the monkey's hand was positioned when the stimulation began. If the hand was positioned above the monkey's head, this stimulation would initiate a downward hand movement. But if the hand was initially in a low position, stimulation of the same site would make the hand move up. Regardless of where the monkey's hand started out, stimulation of the site would bring the hand to the same end position in space relative to the animal's body. This observation yielded an important clue: the movement map does not represent individual muscle movements or even specific limb movements. Instead, it represents goals and target postures relative to the space around one's body. In essence, it is neither a map of action nor of space, but rather a map of action space.

The scientists went on to show that, like so many brain maps, the map of action space is distorted by magnification. When monkeys move naturally, of their own volition, they do nearly all their grasping or manipulating of objects out in front of their chest or face, rather than elsewhere in the space around the body. Likewise, the region of the monkey's reach map representing action space out in front of the chest and head is disproportionately large compared to regions representing other parts of action space. In fact, about three-quarters of the complex hand movements the scientists triggered by stimulating the motor cortex included movement of the monkey's hand toward

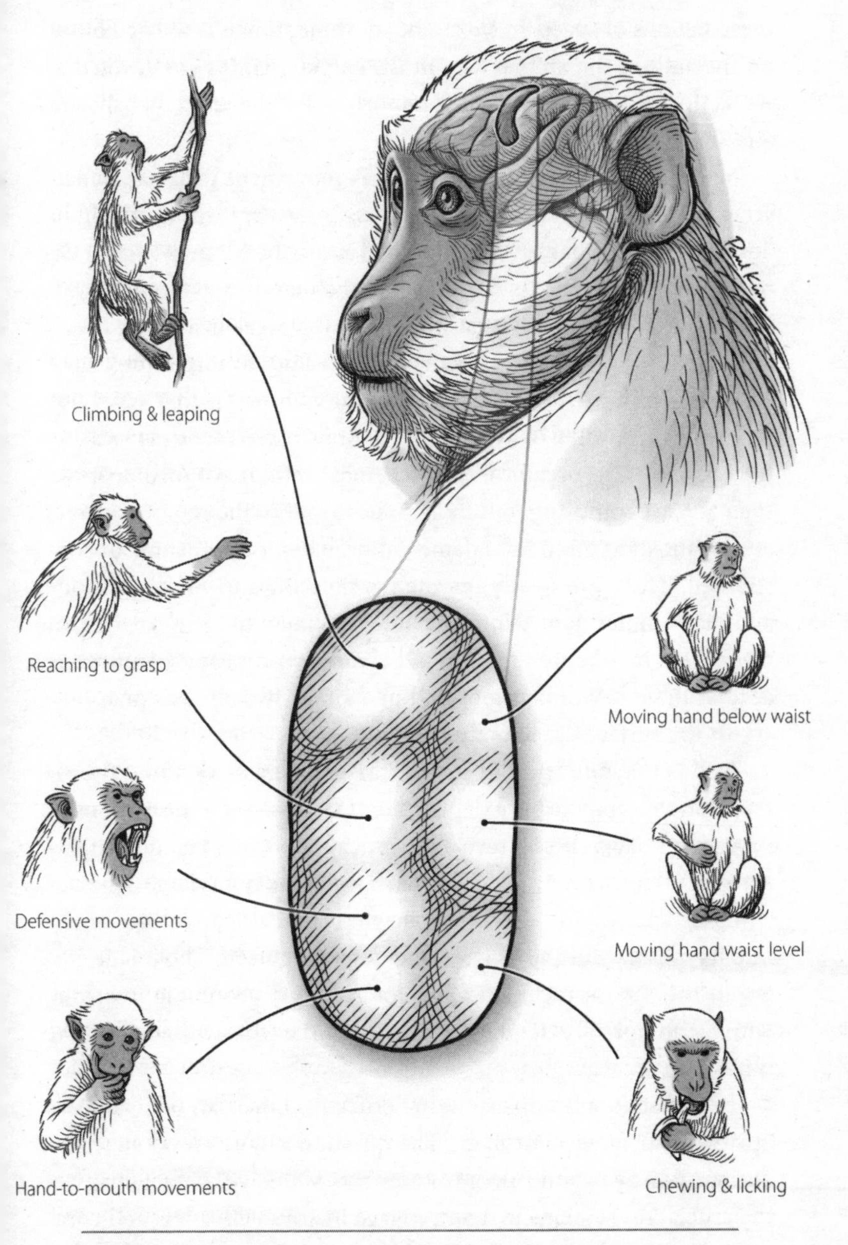

FIGURE 27. An illustration of the zones within the macaque's M1 movement map. ***Paul Kim***

these regions of space. In short, the movement map is warped based on the actions the animal uses most. Larger terrains are devoted to the action spaces that are most important for the animal's behavior and survival.

Scientists' understanding of the M1 movement map has undergone several revisions since Jackson's nineteenth-century clinic. Taken together, what have they learned about the M1 map and the dimensions that dictate its layout? First, the map is organized according to parts of the body that move, but that organization is messy and inexact. Second, it is organized according to target regions of space around a person's body, but only sometimes. Third, it is divvied into zones according to key types of actions required for survival. At first glance, these organizational schemes seem, if not incompatible, then at least confusing. But in fact, the layout of the motor cortex is an elegant solution to a fundamental problem: movement is unconstrained. There are nearly infinite possible ways to move the body and many dimensions that relate one movement to the next. The M1 movement map represents a grand if messy compromise, marrying at least three key dimensions within a single two-dimensional map across the surface of the brain.

Just as magnification in sensory maps offers clues about the vital information a creature depends on for perception, distortions in movement maps should reveal the crucial ways that creatures interact with their surroundings. The macaque monkey has large zones for climbing and leaping, reaching, chewing and licking, putting things into its mouth, shifting its gaze, and defending itself. The motor cortex in the mouse includes zones for running, grooming, reaching with the forepaws, extending or retracting the whiskers, and making ultrasonic vocalizations.

But what about humans? What crucial actions can be read from our own M1 movement map? The question might sound superfluous: wouldn't you and I already know how we move? In fact, many of the crucial movements that you engage in are so effortless and common that you may not consider them movements at all. When you shift your gaze from word to word across the page, it is easy not to realize that you are using ocular muscles to quickly and precisely rotate your eyeballs. When you call out to an acquaintance, you may not consider the remarkable dance of tongue, jaw, larynx, and lips that generates each speech sound. When you reach for a pen to sign your name, you probably do not stop to marvel at how your fingers aim for the object, how they splay just the right amount to grasp it, or how they guide the pen across the paper with exquisite control. In general, we think in terms of our *intentions,* or what we mean to say, use, or accomplish at any given moment. The physical acts we employ to realize those intentions remain more or less unseen and unexamined.

Given that eating is universally important for human survival, self-feeding would be a key candidate for overrepresentation in our M1 map. Like other primates, we use our hands to eat, grasping the food or a utensil, lifting it, and guiding it to our open mouth. Bringing the hand to the mouth is one of the few complex motor actions that human infants are able to do at birth. There is every reason to believe that a sizable swath of the human motor cortex is involved with generating hand-to-mouth movements for self-feeding.

A modern surgical team working with patients undergoing surgery to treat seizures has looked for evidence of hand-to-mouth movements while directly stimulating a patient's motor cortex. The surgeon used long stimulation times (around 1 to 3 seconds) in the hopes of observing complete motor acts rather than twitches only. And that's precisely what was observed: coordinated movements in which the patient's hand closed, the mouth opened, and the arm bent to bring the hand to the waiting mouth. Although the surgeon had limited time to probe and stimulate the brain, the team did observe a few other complex movements triggered by stimulating M1. In one case, a patient's forearm rotated, the fingers extended, and the hand lifted to the level of the head, as if the patient were shielding the face from an approaching object. In another instance, a patient's wrist flexed and the fingers formed a fist, as if grabbing something. In still another, the wrist flexed and the fingers and thumb pressed against one another to form what is called a precision grip: the proper grip for picking up something small, like a pebble.

In this study, it was humans who carried out these elaborate, involuntary movements. Yet their actions closely resembled ones that Graziano and colleagues evoked in monkeys. Humans and monkeys are close evolutionary relatives and share similar bodily layouts—most notably, our dexterous hands and opposable thumbs. Both humans and monkeys bring food to their lips, rather than their face to the food, as most other animals do. We both use our hands for reaching, grasping, and defending ourselves. And if these movements excavated from the M1 movement map in patients are any indication, we may have zones for these actions, as monkeys do.

However, in other ways monkeys and humans move quite differently. There is probably no better example of this than speech. Nonhuman primates vocalize and convey information to each other through their calls. Yet humans are anatomically, cognitively, and culturally endowed with the ability to generate a remarkable variety of precise speech sounds. Crafting them requires great skill and precision. It depends on control of the timing and force of exhalations, on coordination of the larynx, and on the timing and frequency at which your vocal folds open and close, generating the voiced sounds that vibrate your throat when you speak. Producing speech sounds also depends on rapid, precise movements of the tongue, lips, and jaw to sculpt the exiting air into complex patterns of vibrations, which a listener will perceive as vowels and consonants. The spoken word is a miracle of physics, anatomy, and perception, but it is perhaps most of all a feat of motor control. You are probably not a record-setting athlete, an acrobat, or a neurosurgeon, but in a sense, your ability to count aloud from one to ten may outshine the physical feats those professionals can manage. The human capacity to speak may be common, but nonetheless it is spectacular.

It has long been known that humans have an expanded region of the movement map that is involved in producing speech. Although speech requires movement of muscles on both the right and left sides of the body, this region of the cortex is located only in the left hemisphere in most people. It includes the area at the base of the left motor cortex, which controls vocalizations and movements of the mouth. One group of researchers hoped to learn more about motor representations for speech by recording neural activity directly from the brains of patients who suffer from seizures. As in the stimulation studies, there was a clinical need to place electrodes on the surface of each patient's brain before surgery. But unlike the other studies described in this chapter, in this one, scientists did not send electricity through the electrodes and then observe the effect on the patient. Rather, they used the electrodes to listen in on what the neurons beneath were doing. The eavesdropping electrodes formed a grid over the face and mouth portions of the M1 movement map and the neighboring S1 touch map. The researchers asked each patient to read syllables such as "pa" while the electrodes recorded neural activity. Later, the scientists analyzed the recordings to see whether different parts of the cortex were active when participants made different speech sounds.

What they found was a map of the parts of the vocal tract that are used to make speech sounds. A sketch of the map is shown in Figure 28. Activity lower on the map tended to occur when the patients made speech sounds requiring the tongue or larynx. Above that was a zone linked to sounds that require jaw use, and still above that, movements of the lips, as well as a second zone for sounds requiring use of the larynx. In short, this region of the cortex contains its own map of speech sounds based on the body parts we use to create those sounds. And yet this map too is messy and includes some contradictions, such as the existence of two zones to represent sounds articulated with the larynx. It may be that body parts used to make these sounds comprise just one of the dimensions governing the overall layout of the speech map, just as body parts involved in making other movements are not the only dimension governing the overall layout of the M1 map.

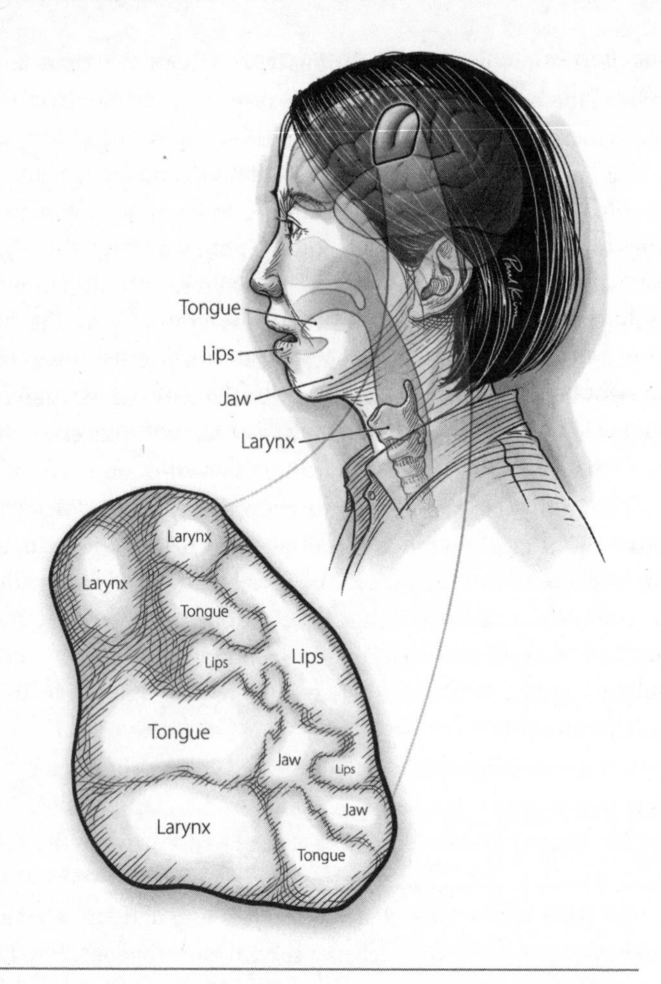

FIGURE 28. An illustration of the vocal tract and the corresponding speech articulator map found in the human touch and motor cortex. ***Paul Kim***

Although the motor cortex in the frontal lobe contains maps crucial to generating action, it is by no means the only residence for maps related to movement in the human brain. Some of these maps make the seemingly impossible aspects of action not only possible, but effortless. In the process, they offer a surprising view of how our brains transform perception into action.

MAPS OF INTENTION

The elderly woman was sitting up in her hospital bed, surrounded by cords, IV bags, and the sound of nurses talking in the hall. Over the IV lines, her wrists were wrapped in bandages. She was recovering from a stroke, but she was alert and determined. What she wanted to do right now was blow her nose. She held a tissue in her right hand, but her left hand grabbed it and pulled hard in the opposite direction.

"What happened to my arm?" she exclaimed.

The two hands fought over the tissue. With her right hand, the patient tried to lift the object out of reach of her left hand, but still the left hand pursued it.

"But . . ." She clucked at her left arm, exasperated. "That's my tissue. Let go of it!" The two hands continued to fight over the prize, pulling in opposite directions.

Half astonished, half frustrated, the woman shouted, "Let go, I tell you!" But the left hand did no such thing. It gave a hard yank, tearing the tissue in two.

The patient held her right hand up and to the side, trying to preserve what was left of her tissue, even while the left hand made several grabs at it. She drew in a couple of quick breaths and looked at her left arm as if she had never seen it before. For all the ninety-four years of her life, both of her arms had carried out her will. "Whatever happened to that arm?" she said. The left hand—persistent, disobedient—paid her no heed.

She used her right hand, still clutching the tissue, to push her left hand out of the way. Once, twice, she shoved her left arm down to her left side, but it popped back up again and grabbed at the tissue. She tried again, two more times, pushing her left arm down so forcefully, it banged against the side rail of her hospital bed. Each time, the left arm just sprang right back.

"How is this possible?" she cried.

The strange behavior of the woman's arm was possible because the stroke had ravaged much of her right parietal cortex, the swath of brain at the very top of the back of the head. Her condition is known as alien hand syndrome because, according to many patients, it feels as if someone or something else has taken control of a hand. The woman described the experience to her doctor: "It feels like a stranger has taken control over my arm and it's doing things beyond my control. But it doesn't obey me, not at all!"

Given that the woman's surreal and maddening condition was the product of parietal damage, we can surmise that the parietal cortex houses brain areas that are vital to action. But what kind, exactly? What system or representation, when broken, could possibly result in such a strange state of affairs?

In considering this question, it helps to step back and appreciate how truly challenging your every action ought to be. Movement in the real world requires integrating information from multiple senses, not to mention data about your moment-to-moment body position and posture. This is no easy feat because your body and senses have many different coordinate systems. Imagine that you see a ball hurtling toward your face. You collect visual information about the ball with a retinal coordinate system. You might also collect touch information about the ball from air movement or the coolness of the shadow it casts on your skin; that information would be in a body surface coordinate system. In order to combine both sources of information and better locate the looming object in space, you need to translate between retinal and face surface coordinates. But the relationship between these two coordinate systems changes every time you rotate your eyes to shift your gaze.

The problem becomes harder still if you need to reach out and catch the incoming ball. Even if you know where the ball is relative to your eyes and your face, that doesn't tell you where the ball is relative to your *hand*. In order to figure that out, you need to know both how your head is angled relative to your torso *and* how your arm and hand are positioned relative to your torso. When you add up the computations involved in simply catching a ball, they seem astronomical. And yet to have any hope of catching it, we have to perform them all in a fraction of a second. Similarly complex computations are involved in steering a car or defending yourself from attack.

The parietal lobe is home to several crucial maps involved in generating movements and solving this conundrum. These maps are not purely motor or sensory; they actually combine and align information from touch, vision, and hearing with information about body position and space around the body to which actions might be guided. For example, there is a parietal zone in monkeys and humans involved in defense of the face. There, information about touch sensations on the face is combined with visual information about objects close to the face, so that the area contains coexisting maps of the face (for touch information) and the retina (for visual information). If a ball came hurtling toward your nose, this map is well equipped to determine the direction from which the ball is approaching and to help prepare directions to the motor cortex about where and how to duck or shield your face.

Many other wonderful maps populate your parietal cortex. For example, a touch map of the body, skipping the face and hands, overlaps with a visual map. These maps combine sensibly, so that touch information from your legs and feet overlaps with visual information from the lower part of your visual field, below where you are directing your gaze. This combination makes sense because objects detected in your lower visual field will tend to be low to the ground, near your legs or feet. Likewise, objects that touch your legs or feet will probably initially be present in your lower visual field, at least until you shift your gaze to look directly at the culprit. Other parietal regions are devoted to the hand and arm. One such region supports reaching and another, grasping, while the region between them appears to be involved in both. The reach area, for example, must align visual information about an object's location with movement plans for the physical act of reaching.

What does it actually mean to have overlapping maps from multiple senses? The simplest answer is that the neurons within the map are receiving and responding to information from more than one sense. But they can do so in a variety of ways. Some parietal neurons have receptive fields that jointly reflect inputs from more than one coordinate system. Others have a receptive field in one coordinate system (say, spatial location relative to your direction of gaze), but the intensity of these neurons' response can be amped up or tamped down by information from another coordinate system, such as where the hand is currently positioned. Thanks to the unique receptive fields and connections of neurons in these parietal maps, they are able to combine and align disparate bits of information about different parts of the body. The combined effort of these neurons in their specialized maps solves the challenging computations of catching a ball that is hurtling toward you by making such computations lightning quick, effortless, and more or less accurate.

To appreciate the importance of integrating these modalities and coordinates, you need only consider the kind of impairments people experience after the parietal cortex is injured. Damage to one area leaves patients unable to use visual information to guide movements of the arm or hand; these patients are able to see and name an object before them, and yet they cannot reach out and grab it. Damage to other areas interferes with patients' ability to shift their gaze toward a target or to keep track of their own limbs in space. Still other patients lose awareness of all space—and everything in it—on one side of the body.

The complex yet essential maps in the parietal cortex seem to defy any single pithy label. They have been described as visual maps, multisensory maps, spatial attention maps, and movement maps. You could consider them as spatial maps representing the locations of behaviorally relevant objects around you. In general, those objects and people to which you direct your attention are also the objects and people that you are most likely to act upon or interact with. Paying attention to an object in space allows you to monitor it and to prepare specific goal-directed actions toward or away from it, as needed. If I am driving my car beside a bicyclist on the road, I will pay careful attention to her as I pass. Paying attention allows me to react if she swerves suddenly. Attending to her in relation to the space around me allows me to quickly determine *how* to react—for example, how I should navigate my car to avoid colliding with her. In short, attention highlights regions of space for potential action.

Some researchers have described these parietal maps as maps of intention, in that they represent the locations to which we will likely make object-guided actions. Ultimately, there may be very little daylight between these concepts. The difference between potential action, intended action, and the initiation of physical action may actually be a matter of degree, at least as far as representation in the parietal cortex is concerned.

We do know that the parietal cortex does not directly make movement. Instead, it sends signals to the motor cortex in the frontal lobe, which in turn generates actual physical movements. Scientists have studied this by stimulating parts of the parietal cortex in animals, which would normally cause them to make a movement. But if the scientists used chemicals or temperature to temporarily shut down the animal's motor cortex, stimulating its parietal cortex no longer would cause the animal to move.

All of this can help us understand the trials of the stroke patient with a disobedient left arm. The symptoms of alien hand syndrome are astonishing and disorienting because the affected hand appears to be acting with *intent,* but not the intent of the person to whom the hand is attached. The aftermath of the woman's stroke would seem far less bizarre if the arm were paralyzed, weakened, or spastic. But the arm moves in a coordinated and precise fashion toward a goal—in this case, the woman's tissue.

Although the parietal cortex aligns and relays information about targets for possible actions, it can unlock physical movements only by way of the motor cortex. If the right parietal cortex is damaged, then the right motor cortex will not receive proper information about intended actions in space. Evidence from another patient suggests that the motor cortex goes rogue in the case of alien hand syndrome. Lacking the right parietal lobe's control of intended actions, the right motor cortex may receive incomplete information from other parts of the brain, thereby triggering faulty actions. In the case of our patient, her intention was to grasp the tissue with her right hand, so her left parietal cortex correctly sent a signal to her left motor cortex, triggering the action with her right hand. Yet her right motor cortex, deprived of guidance from her damaged right parietal cortex, followed the same instruction, causing the left hand to reach for the tissue as well. The result was a war of intentions—two hands controlled by two halves of the brain, one functioning as it should and the other cut off from guidance.

Several themes emerge from the maps and zones in the motor and parietal cortex. The brain's representation of movement is ruled by many dimensions: what kind of movement it is, which parts of the body are moving, and where in space you move, toward or away from something. Magnification reveals the outsized importance of key body parts, such as hands, which we rely on most to carry out our will. Most of all, the movement maps in your brain demonstrate the paramount importance of an action's goal. You move for a purpose, whether it is seizing a reward, fleeing an enemy, or reaching a destination, and your survival may rest on the results. All the rest of it—the coordinate systems, alignments, and countless computations—is simply built into your maps, running in the background and silently, invisibly, making each of your actions possible.

Do our movement maps develop as they do *because* we move the way we do, or do we move the way we do because of our preexisting movement maps? More broadly, do our maps make us, or do we make our maps? The answer to these questions reveals the role of early life experiences and environments in brain development and the warping of brain maps. The results can have lifelong effects on a person's health and ability to perceive or move . . . for better or for worse.

7

Maps in the Making: How Brain Maps Develop and Adapt

YOUR BRAIN MAPS BEGAN when you were just a speck of cells folding in upon itself like wet origami. Cells divided, shifted, clustered, and took on unique identities as ancestors of the various parts of the body you would one day possess. Your nascent brain bubbled up at the tip of your nascent spinal cord. Neural structures began to form. Newborn cells left their neural nursery and traveled to the proto-regions that would become their home. Their migrations laid the basic blueprint for the zones of what is now your brain, specifying the relative size and location of areas like V1 and offering early clues that yours would be a human brain and not that of a turtle or a whale. In that warm, dark stew of primordial you, these cells navigated just as bacteria navigate toward nutrients or sperm navigate toward an unfertilized ovum: by sensing a chemical that attracts them and automatically moving toward that chemical. Altogether, these cells carried out an exquisite choreography orchestrated by your genes.

Once neurons settled into their new homes, they sprouted axon stubs that bravely and blindly inched toward unseen targets. These axons too were steered by chemicals that seeped out from landmarks in your tiny brain-ball and either attracted or repelled them. The axons were picky little fellows, seeking just the right balance of chemicals, which equated to the proper target location within your developing brain. Among these pilgrimages were axons journeying from the structure that would become your eye to the structure that would become your thalamus, an important brain area that lies beneath the mantle of the cortex. The gradients of seeped chemicals provided enough information to ensure that the axons of neighboring cells in the retina of each of your developing eyes landed upon neighboring cells in the part of the thalamus that would be devoted to vision. In turn, neighboring cells in the visual part of the thalamus sprouted axons that journeyed to your future V1 and created connections with neighboring cells there. Through this pattern of connections, a map of your retinas could be re-created in the visual thalamus and then again, in V1. Many such pilgrimages were taking place at the same time between incipient regions of your brain. The result? Before you had functioning eyes, your proto-brain already contained the structural scaffold of your visual map in V1—and many other brain maps besides.

When axons from your developing thalamus reached their assorted targets in your cortex, they brought signals from the sensory receptors in your eyes, ears, tongue, and skin to the cortex for the very first time. Not that there was much to report. Your amniotic surroundings muted much of the world beyond. Once the basic wires for your maps of sensation were in place in your brain, your body assumed the role of tutor. The cells in the developing retinas of your eyes began generating slow, spontaneous waves of neural activity. In terms of the typical speed of brain activity, these waves were positively glacial, occurring only once every minute or two. Because neighboring cells in the retina connected to neighboring cells in the thalamus, which in turn connected to neighboring cells in V1, the waves in the retina naturally traveled through these connections and created parallel waves in the visual part of the thalamus and in V1. The parallel and nearly simultaneous waves in your visual areas refined each visual map and strengthened their alignment, all before your eyes ever opened.

Your auditory maps were tutored in a similar way. A temporary structure in the fetal cochlea triggered spontaneous waves of activity in your sound receptors, which sent the waves on to the auditory zone of your thalamus and then on to the A1 sound-frequency map. These waves refined your auditory maps long before you left the waters of the womb.

Tutoring your touch maps proved a little more challenging. The touch receptors embedded in your waxy fetal skin were sprinkled all over the surfaces of your body and could not, like the cells in the retina and cochlea, generate a coordinated wave. Instead, your spinal cord triggered random, rapid twitches of your arms and legs. These fetal twitches caused the moving body part to brush against the amniotic sac and uterine walls and, in doing so, created waves of pressure across the surfaces of your skin, which activated your touch receptors. These waves of activity spread to the region of the thalamus responsible for touch. From there, it continued to S1, refining the body-surface layout of its touch map. But the impact of twitches didn't stop there—S1 sent these waves right on over to its next-door neighbor, the swath of cortex that would become your M1 movement map. In doing so, the touch map was refining the underlying body representation in your future movement map.

Up to now, I have described S1 and M1 as though one is purely tactile and the other related to movement. Although they are often presented in this way, in truth they are not so easily separated. Even in the adult brain, one can find what appear to be neural responses representing movement in S1 and what appear to be touch responses in M1. We might call S1 mostly tactile and M1 mostly movement-related. Or, as many scientists do, we could refer to them jointly as the sensorimotor cortex. It may be that the mature S1 and M1 have these interlacing roles because movements and tactile sensations are often linked in our everyday life outside the womb. Moving your body generates feelings on your skin, and often, sensing something on your skin causes you to make a movement. But it might also be a vestige of the uterine beginnings of these maps, when touch cortex taught movement cortex about the body it would command.

As you neared the end of your third trimester of gestation, all the major highways of connections in your brain had formed. Your cortex had wrinkled up like a raisin so that it could grow five times larger while still fitting within the soft, shifting confines of your skull. You had brain maps ready and waiting to process the world you would soon be born into. And yet you had not yet taken a breath of air, supported the weight of your body, or had your first glimpse of direct sunlight. Does that mean that your brain maps today were irrevocably shaped before birth by your genes? Not at all. Your genes took the first remarkable step. From metaphorical clay, they sculpted an intricate scaffold upon which the final product would be built. They provided the raw materials and the basic structure of your brain and its maps. But as soon as this structure was in place, your brain began to sense and learn from your environment. This powerful period of learning had the potential to refine or even rearrange your brain maps in ways that would affect you for a lifetime.

LEARNING ABOUT THE WORLD WHILE YOU ENTER IT

Your time in the uterus offered some early clues about the world into which you would be born. You could at least partially hear your mother's voice, which contained information both about her and about the language that you would one day learn to speak and understand. The rhythm of her heartbeat and the way that her movements jostled you would have been important features of your fetal world. And with your birth, you were ushered into a world teeming with sensation. Your sensory maps, newly wired by genes and trained by waves, were now bombarded with clues about what your new world contained and how you could harness its information to greatest effect.

One way scientists study the impact of early experience on brain maps is by manipulating the environments or sensory experiences of newborn animals. There is more than one way to alter an infant's environment. You can *add* something to it that the infant would not normally be exposed to, you can *remove* something that would normally be present, or you can do *both* at once. Studies from animals show that either adding or subtracting from a newborn animal's normal environment can remodel its brain maps, with long-term repercussions.

Many of the studies investigating the impact of early experience on brain maps have focused on sound-frequency maps, like A1, in the auditory cortex of rats. Unlike human infants, newborn rat pups cannot hear or see until more than a week after birth. When the pups are born, they have a rough frequency map in A1 ranging from about 1,000 to 32,000 hertz, with scant room devoted to frequencies higher than 32,000 hertz.

Scientists have identified several types of ultrasonic rat calls, including distress calls at 22,000 hertz, made when they are in pain or detect a predator, and joyful calls at around 50,000 hertz, made when they are playing, mating, or being tickled. The sound of the distress call causes other rats to freeze or flee, whereas the joyful calls beckon fellow rats to approach and trigger activity in reward centers of their brains. Like laughter for humans, the 50,000-hertz calls foster social interactions and strengthen bonding.

When rat pups begin to hear the sounds of the world around them, about twelve days after birth, these sounds will naturally include the joyful calls, at 50,000 hertz, from their mother and other rats in their colony. Although the newborn A1 map spares little room for frequencies above 32,000 hertz, after two weeks of hearing its environment, the rat pup's A1 map is remodeled, so that some of the territory that processed sounds between 20,000 and 32,000 hertz now processes sounds above 32,000 hertz. In fact, about 40 percent of a rat pup's A1 terrain is now devoted to processing sounds between 32,000 and 64,000 hertz.

But this rapid redistricting depends on experience. When scientists blocked the pups' ears for two weeks after hearing would normally begin, their A1 maps remained organized according to sound frequency, but they devoted far more territory to processing sounds around 25,000 hertz and far less to processing frequencies around 55,000 hertz than was the case for the maps of their hearing littermates. The illustration in Figure 29 depicts this difference. In short, those pups that heard the crucial 50,000-hertz calls in the first few weeks of life had A1 maps that shifted to magnify that frequency range. But if a pup's auditory experience lacked this normal input at this crucial time after birth, its A1 map would not invest the usual amount of brain space to representing sounds at that frequency. The

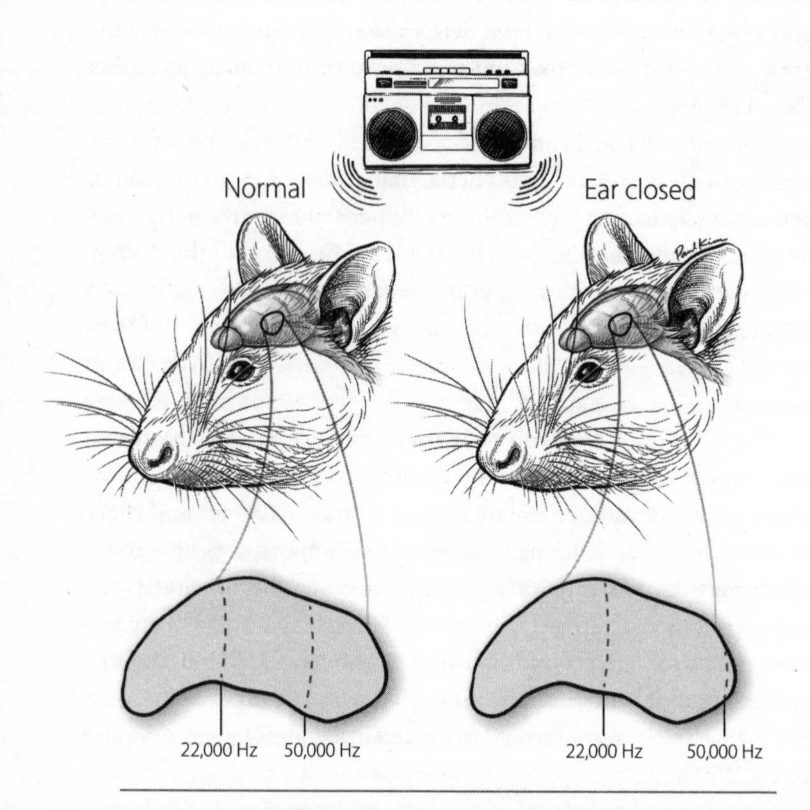

FIGURE 29. The effects of early experience on the development of the A1 sound-frequency map in rats. ***Paul Kim***

experiment did not go on to test how the pups with altered A1 maps might have fared if their hearing was restored. But if returned to a normal life, these pups would have begun that life with brains optimized for hearing squeals of fear yet unprepared to hear the playful trill of a companion or a mate.

Other experiments have shown that adding unusual new sounds to a newborn animal's environment can also remake its A1 map. If rat pups are placed in a sound chamber and exposed to either a pulsed low-frequency tone (4,000 hertz) or a higher-frequency tone (19,000 hertz) for the first few weeks of life, their A1 maps will be remodeled so that more of the map represents tones around 4,000 hertz or 19,000 hertz, respectively. Even if the pups are returned to a normal sound environment after three weeks and raised normally thereafter, their A1 maps in adulthood will still be dominated by the frequency that they heard in infancy. Exposing an adult rat to the same monotone pulses has no such effect on its A1 map. There is something special about these first few weeks of hearing, when the layout of the rat's A1 map is particularly malleable and influenced by incoming sounds. In effect, this brief window gives creatures the opportunity to sample inputs from their environment and revamp their brain maps accordingly. It gives each newborn creature the chance to adapt to its specific environment, allowing it to specialize in processing and perceiving the types of sounds that it will presumably hear for a lifetime.

The A1 maps of these rat pups offer a glimpse of a much grander and more sweeping phenomenon, which encompasses maps of all the continuous senses in animals ranging from birds to cats, sheep to toads, fruit flies to humans. Early in life, brain maps learn from a creature's environment and are remodeled by it. A nascent map laid out according to a genetic blueprint is first tutored by the body and then intensively trained by the wider world. Learning shifts the circuits and connections bringing information into and out of the brain map. As a result, the neurons within the map change their receptive fields. Whereas a patch of neurons within the twelve-day-old rat's A1 map fired fastest to sounds around 30,000 hertz, the same neurons in that patch responded most to frequencies around 45,000 hertz in the twenty-two-day-old rat. As neurons in whole regions of a map change their receptive fields, the overall map is remodeled and warps. Areas representing crucial pieces of the world (say, 50,000 hertz for A1 or foveal inputs for V1) expand their estates. There are winners and there are losers. For example, magnifying the fingertip representation in S1 might involve stealing territory from the S1 face representation.

Changing magnification in a brain map by shifting cells' receptive fields is just one way that the developing brain invests resources according to experience early in life. Another is investing in the materials that make up the map itself. Maps are generally thought of as flat representations with two dimensions: a length and a width. But of course maps, whether in the brain or elsewhere, exist in a three-dimensional world. For example, if you hold a printed land map in your hands, that map has a third dimension too: the thickness of the paper and the ink that make up the map. In the case of printed maps, this material thickness might affect how durable or legible the map is, but it has no bearing on the information the map represents.

Brain maps are made of materials too: neurons and their wiry extensions, the axons and dendrites that transmit messages between them. Another way that developing brains can learn and invest is by altering these materials in ways that make key regions of the map *thicker*. For example, if your caregiver often tickled your toes or played games like "this little piggie" when you were a baby, these early enriching experiences could have caused the toe zones of your S1 touch map to *expand* (so that it took up a greater portion of your brain surface), to *thicken*, or both. When learning results in the thickening of part of a brain map, that generally means that this part has *more stuff*—more neurons, more connections between neurons, or more support cells to help those neurons do their job. If the left-toe regions of your S1 map are thicker, it can indicate that the neurons representing touch on those toes are better equipped to communicate and coordinate with one another. The result is better local processing, which should make you better at detecting or discriminating between touches on your toes. In childhood and, to a lesser degree, later in life, brain maps can invest and adapt by magnifying key territories of the map, by bolstering these territories with material support, or by doing both at once.

It is one thing to consider how experimental manipulations cause rat pups' brain maps to be remodeled. But the same principles apply to human infants and the neural impact of their early environments. Consider infants who are born extremely prematurely and require extensive medical care and hospitalization after birth. These infants can spend weeks to months in neonatal intensive-care units, where they are often exposed to the loud, high-frequency noises of ventilators, fans, infusion pumps, monitors, and alarms—additions to the auditory environment they would typically have experienced in the womb. But there are also subtractions to consider. If these infants were still in the womb, gestating until they were full term, they would experience an acoustic world dominated by the melodic low-frequency components of their mother's voice and the predictable rhythm of her heartbeat. These sounds are all but absent in the premature infant's hospital environment. There are other environmental differences as well. For example, hospitalized infants tend to experience more painful procedures, such as blood draws, and less person-to-person physical contact than they would receive either in the womb or at home with their parents.

Although we don't yet fully know how hospitalization affects babies' developing brain maps, evidence is accumulating that it impacts the development of sound and touch processing. Of course, the care these infants receive during their stay at the hospital is essential to their survival. But a growing scientific and medical awareness of the importance of early life environments on neural development has caused hospitals to reexamine the sensory environments in these neonatal intensive-care units. Can they provide lifesaving care and at the same time increase babies' exposure to social touch? Can they reduce the noise caused by alarms and equipment? Many hospitals have started working to address these important concerns.

In some cases, meaningful changes might be surprisingly easy to make. For example, one study found that simply playing an audio recording of the mother's heartbeat and the low-frequency components of her voice (that is, the sounds that would have been audible to the infant in the womb) for three hours per day while a premature infant was hospitalized affected development of the auditory cortex. The researchers found that this part of the brain, which includes the A1 frequency map, was thicker in those babies who heard these recordings compared to those who did not.

The profound cognitive and perceptual impact of early-life sensory experiences affects all infants, not just those in hospital care. Another example concerns babies born with cataracts, or cloudy deposits that block vision in one or both eyes. These deposits deprive the eye and the visual brain maps like V1 from receiving patterned visual inputs after the baby is born. Research both in humans and animals has shown that this deprivation wreaks havoc on the V1 map and other visual maps in the brain. These maps develop based on the blurry or one-sided inputs they receive. If a baby's cataracts are surgically removed within the first few weeks after birth, he or she has a good chance of seeing more or less normally as an adult. But for every week that surgery is delayed, the odds that the child will eventually recover normal vision go down. Adults who had cataracts as infants tend to be impaired at seeing fine visual details, detecting certain complex movements, and recognizing faces. These challenges do not stem from anything that is happening in their eyes, but instead from what has already happened in their brains.

Examples like these illustrate how sensitive brain maps are to environmental factors in the earliest days of life. Even seemingly minor disruptions at that time can have lifelong repercussions on brain maps and sensory perception. You could be forgiven for thinking this early vulnerability sounds strange, or even nonsensical. Given the many ways that our bodies and brains are molded by our genes and the evolutionary forces at work on them, it is surprising, even vexing, to think that we should be so reliant upon mere sounds, sights, and sensations. But in truth, this strange vulnerability is also our superpower, because it allows each and every creature the chance to uniquely and responsively adapt to its surroundings.

In the context of evolution, creatures adapt to an environment because genes that help them survive and procreate in that environment are more likely to be passed on to offspring. This kind of adaptation happens over generations. But I am talking about a different kind of adaptation, one that happens uniquely to every individual brain over a period of days, weeks, or months. While you were in the womb, and during those bright, heady days shortly after, your brain was gathering clues about the world where you would live. Based on these clues, your brain maps became remodeled and redistricted, investing your finite neural resources in processing the specific kinds of sights, sounds, and touches you received, at the expense of processing others that you did not. By and large, this kind of adaptation should be a great investment. In the natural world, environments tend to be relatively stable over short timescales. The environment into which you are born is likely to be, more or less, the environment in which you will die. Given all of the possible environments in which you might live out your life, it makes sense to invest and prepare for life in the environment you sensed as an infant. It is only when this general rule fails, as in the case of a child who begins life in intensive care or temporarily blinded by cataracts, that this form of adaptation works against us.

To see how this adaptation can help rather than hinder, consider an infant born with retinas that do not function. This can happen for various reasons, including inherited disorders and extremely premature birth. Unlike cataracts, these retinal conditions cannot be corrected with surgery. So this child will be blind at birth and presumably for a lifetime. She has no use for a visual cortex, complete with its intricate V1 map. If she were not able to adapt to her circumstances, she would be left with a large chunk of brain that would remain useless. This would be bad enough because, as we've seen, brain space is anything but cheap. But beyond that, such a child would face particular challenges, from finding her way around to identifying her peers and avoiding dangers. She would also have to rely on her other senses for information, such as reading text with her fingertips and locating objects like cars based on the sound they make. She could not afford to leave a chunk of her brain unused.

Thankfully, she would not grow up with part of her brain lying idle. Even before her birth, her brain might be changed by her blindness. She may lack the retinal waves that normally sculpt visual maps. The absence of normal visual activity before birth and of normal visual inputs after birth will drive changes not only in her visual maps, but also in her maps for touch and hearing. Ultimately, blind children go on to radically redistrict their visual cortex. For example, the region at the back of the brain that would typically contain the V1 visual map instead supports language processing in people who are blind from birth. In this way, areas of the brain that would typically process visual inputs do not go to waste in a blind child's brain. She can harness these areas to process other crucial inputs, like language. In short, the radical rezoning of the "visual" cortex gives a child born blind the opportunity to adapt to circumstances and put her brain to the greatest use in service of her particular situation and needs.

Redistricting the "visual" cortex helps a brain adapt to life without vision, but these benefits are not equally enjoyed by all. The "visual" cortex of adults who have been blind since birth or since very early childhood is much more dramatically remapped than those of adults who become blind later in life. The important difference appears to be how old the person was when she became blind, rather than how many years have elapsed since then. Once a child's visual cortex has begun the work of processing visual information and crystallizing its visual maps, there are limits to how much this territory can be transformed later on.

This is a general theme across brain maps: infancy and, to a lesser degree, childhood are particularly active periods for sculpting and refining brain maps. The dramatic reorganizations that are possible in the infant brain simply do not happen in adults. Still, many features of brain maps remain somewhat malleable, enabling adult animals to fine-tune their perceptual and motor abilities in response to the demands of their environment. This kind of learning bears some resemblance to the adaptation that occurs in infancy, but it is different in at least two critical ways. First, it permits change and remapping on a substantially smaller scale. And second, it is driven not by mere *exposure* to sights, sounds, and the like, but by the *significance* of these inputs. In other words, it is shaped by the adult animal's goals and behaviors.

To illustrate, consider the rat pups that heard a pulsed tone for the first weeks of life. As a result of this exposure, their A1 maps became warped to magnify the representation of the tone's frequency. These tones were irrelevant. They did not signal that anything was about to happen, yet their presence in the environment was enough to remodel the pups' auditory maps.

Compare that with the results of another experiment in which adult rats were exposed to a frequent 6,000-hertz tone. The study included two groups of thirsty rats: one that heard the tone just before being given a water reward and another that also heard the tone but never when it was receiving its water reward. After the training, only rats in the first group had expanded their representation of frequencies around 6,000 hertz in their A1 maps. Yet both groups of rats showed an expanded region of the map around 1,000 hertz. When the experimenters studied the sounds in the rats' cages, they realized that the water delivery device emitted a low-frequency whir around 1,000 hertz each time it delivered a water reward. Unbeknownst to the experimenters, the rats had learned to detect and capitalize on an additional signal that precious water was forthcoming.

What mattered for the adult rats was how relevant the sound was to them. This kind of learning is probably driven by reward and stress signals in the brain—spurts of neurotransmitters that are released when something good (like food or sex) or something bad (like pain or threat) happens. The name of the game is identifying signals in your environment that occur just before a good or bad event. If your brain maps can adjust and invest to better detect these signals, you can get a head start on predicting and reacting to whatever good or bad things may come your way.

Although people can learn from their environment throughout life, childhood and particularly infancy are periods of astonishing neural flexibility and adaptation. During this time, brain maps can be remodeled or entirely upended based on a child's sensory experiences. Unlike the learning that takes place in adulthood, changes to brain maps during infancy can be dramatic and long-lasting, and can be triggered by irrelevant sensory exposures. These facts highlight the unique role of early life experience in shaping the layout of brain maps for a lifetime. But this information emerged from studies of development in sensory brain maps like the V1 visual map and the A1 sound-frequency map. What factors determine the layout of our movement maps?

LEARNING TO MOVE

Movement maps could be thought of as the black sheep of brain maps. Although they abide by many of the same principles as sensory brain maps, they play a wholly different role in keeping you alive. Like sensory maps, your M1 movement map was wired up by chemicals and then tutored by your body in the womb. That means you had at least a basic movement map long before you were tying your shoes or writing your name. But what capacity for movement did you have when you were born? And how have your experiences since then shaped the movement maps you rely on today?

A newborn baby's capacity for movement is strikingly unimpressive. Babies cannot move themselves from place to place by scooting, crawling, walking, or swimming. These sad creatures can barely see a foot beyond their nose or generate sufficient body heat. They cannot even lift their own head. In a showdown between an ant and a newborn baby, you would be wise to put your money on the ant. And yet new parents are often a pitiable bunch, run-down and under-slept on account of a baby that literally could not hurt a fly. It would seem that babies are born with a small arsenal of actions that they use to great effect. Perhaps the most powerful of these is crying, something they can do from the moment they are born, and smiling and laughing, which they can do socially within weeks of birth. Human parents find these actions highly motivating. Thus, infants leverage their limited repertoire of movements to compel caregivers to meet their survival needs.

Beyond crying and smiling, newborns can also bring their hands to their mouth and suck their thumb. Bringing the hand to the mouth is a fundamental and early-emerging action represented in both the monkey M1 and the human M1. This action might be written into the scaffold of the developing brain before experience makes its grand entrance. Alternately, it might result from fetal experience. The close proximity of the hand and the mouth may have brought them into frequent contact in the womb, and the tactile thrill of brushing these two sensitive body parts together may have reinforced the action. Whether hand-to-mouth action was the product of programmed cell wiring, fetal experience, or both, it beautifully prepares the infant for the crucial lifelong activity of feeding itself.

A host of other actions round out the newborn's repertoire and form the building blocks of more complex movements to be mastered later. On daily display are the baby's lip smacks and puckers, opening and closing of the mouth, widening of the eyes, furrowing of the brow, grasping of the fingers, kicks, coos, and babbles. Caregivers react to these actions, encouraging some and discouraging others. How they do so depends on many factors, including the caregivers' culture. Culture shapes the social interactions and physical experiences that drive motor learning.

Most of us have witnessed child-rearing only in our own culture, and so it is easy to assume that babies learn to move in the same way all over the world. This is not the case. For example, if you were raised in a farming community in western Kenya, you would have spent the first few months of your life sitting up in your mother's lap while she drank tea, shelled maize, or talked with adults. Compared to a baby born in America, you spent far more time sitting and far less time lying down. Nearly every day your mother played *kitwalse* ("to make jump") with you, holding you under your arms and bouncing you on her lap to trigger your stepping reflex. Building upon your newborn reflexes, she actively taught you to sit, stand, and walk, so that you could do all of these activities about a month sooner than babies raised in America. But because you spent little time lying down, you took longer than your Western counterparts to learn to lift your head and to crawl. By shaping your earliest physical experiences, your mother's daily activities and cultural norms determined how and when you learned even your most basic early actions.

Although children the world over eventually learn to sit up, to walk, and to speak, our early experiences and actions can drive lasting differences in how we use our bodies to achieve physical goals. The activities that we engage in, as children and as adults, determine our repertoire of actions and guide the layout of our mature movement maps. For example, early experience with playing a musical instrument affects the layout of the M1 movement map. In right-handed people, the area of M1 that represents the right hand is typically larger than the one representing the left hand, presumably because they use their right hand to perform more motor acts. However, playing the piano requires precise movements of the fingers of both hands. One study used brain scans to measure the M1 movement map in right-handed piano players and in right-handed people who weren't musicians. Overall, the region of M1 representing the right hand and the region of M1 representing the left hand were both larger in piano players than in non-musicians. But even among the piano players, this effect depended on when musical training began; the hand regions of the M1 maps tended to be larger for musicians who began playing when they were three, four, or five years of age compared with those who began training at around eight, nine, or ten years of age.

The specifics of how brain maps are affected by musical training depends not only on when training begins but also on the specific instrument and how it is played. For example, piano playing requires fine movements of fingers on both hands, whereas playing string instruments like the violin calls for such movements only from the left hand. Whereas piano players have expanded representations of both the left and right hands in M1, the effect is seen for only the left hand in string players.

There is always a risk of misinterpreting results like these. Might people with naturally large M1 hand representations be more likely to take up musical instruments earlier, or stick with them longer? How do we know which came first: the brain map layout or the instrument playing? One experiment dealt with this question by assigning six-year-olds to either take individual weekly piano lessons or participate in group music classes without playing instruments for fifteen months. They also tested the children's motor ability and conducted MRI brain scans both before and after training to see how their brains had changed structurally. There were no differences between the brains or motor abilities of the two groups of children before training began. But after the fifteen months were up, the region of the M1 map representing the left hand had grown larger in kids who had received piano lessons but not in those of the other group. In other words, the kids who received piano training eventually showed brain map patterns like those found in adult pianists. Moreover, the more they showed this brain pattern, the better they performed on a non-musical movement task involving the fingers of the left hand. All of this suggests that adult pianists have large hand regions in their movement maps because they played a lot of piano with their hands, and not the other way around. Their *experience* of using the hand changed their M1 map, promoting better overall dexterity of that hand.

All told, numerous studies support the idea that training with a musical instrument, particularly in childhood, affects the layout of not just movement maps, but also tactile and auditory maps. But that does not mean that the brain maps of child musicians are superior to those of their non-musical peers. Piano lessons do not buy your child a better brain. Instead, those lessons (or rather the hours of practice that they promote) buy you a brain that is better suited to *piano playing* and other tasks that require dexterity of the hands. It does not make you a better soccer player or a better hula-hoop competitor. There is no single right way to train a brain and shape its maps. The key is to use your senses and interact with your environment, to practice skills that you have learned, and to keep learning new skills. If there is something in particular that you need to be able to do, practice doing that. That is how to train young brains and older brains alike.

The studies of the impact of training with a musical instrument illustrate how the activities we engage in mold our movement maps, which allow us to perform well in certain activities. No one would be surprised to learn that practicing a physical action or skill allows you to perform it more easily and fluidly in the future. But consider how this evident transition may stem from unseen changes within your movement maps. Just as stimulation of a monkey's M1 cortex evoked the fluid, complex actions that the monkey typically engaged in, stimulation of your M1 cortex could unveil the meaningful, practiced actions that you make every day.

Scientific studies have not yet provided us with an inventory of human M1 action zones and the specific practiced actions that lie within. We cannot easily study them with functional MRI brain scans because these scans are ruined if people make large movements during scanning. Yet it is reasonable to assume that there is no universal human movement map, just as there is no single set of actions that human beings engage in. Movements that you make day in and day out, particularly early in life, appear to etch themselves into the motor cortex. If we were to open up the skull of a pianist and stimulate the left-hand zone of his M1 map, perhaps we might see the fingers of his left hand play over invisible keys. The same might be true if we examined the right-hand region of M1 for surgeons who suture delicate tissues day in and day out, or for children in sweatshops who labor over detailed embroidery.

It may even be that our movement maps have a grander tale to tell about who we are, where we are, and how we live. Does a person spend hours playing video games, knitting, or praying? Do they traverse their environment by running, driving, bicycling, or paddling a boat? Do they feed themselves with a spoon and fork, with chopsticks, or by scooping their food with bread? Compare the action repertoire of the savanna hunter to an urban hairstylist, or of someone who completes marathons on foot versus in a wheelchair. Action is the ultimate nexus of experience, environment, body, and brain. In some ways, your movement maps will be like that of a monkey or a rat. In others, they may be quite different from those of even your sibling or your spouse. It is a wondrous thing to consider: the very maps that give you the gift of movement have been sculpted by your movements in the past. They bind together past and present, so that the movements you have made before are also those you're best prepared to make again.

The remarkable thing about learning and development is that it simply happens. There is no need for a puppeteer or a ghost in the machine. The developing brain follows physical and biological laws that dictate how it learns from its surroundings. There is no "should" about this kind of learning. No supervisor comes to confirm that this remarkable process of adaptation is in a creature's best interest. It just blindly unfolds to make you and me and everyone else the way that we are.

Scholars have spilled boatloads of ink in debating whether the human mind is forged by nature or nurture. The question itself is ill-conceived. Your genes provide a basic brain structure that is sufficiently complex and changeable to capitalize on the properties of the world into which you are born. Your experience, particularly early in life, provides the information that dramatically teaches and refines that structure. Experience begins in utero as the physicality of your body offers your brain early clues. And yet your genes drive and maintain the neural mechanisms that allow you to learn and adapt throughout life. There is simply no point at which the impact of genes ends and that of experience begins. The two are locked in an elegant dance that began in the womb and does not end until you draw your final breath.

8

Knowing Again: Brain Maps for Recognition

You were born equipped with burgeoning visual maps, which matured in the months that followed. These fortuitous events allowed you to see. But was seeing enough to throw open the doors of meaningful perception? Was it enough to unlock essential information about where you were and who or what was around you? Absolutely not. Deciphering this information requires something else. We know that because we can observe what happens when brain damage steals that "something else" away.

John was one of the countless people who have lost this mysterious capacity. He lived an ordinary life without visual problems until middle age, when he suffered a stroke after appendix surgery. The stroke ravaged large regions on both sides of his occipital and temporal lobes – regions outside the V1 visual map but not far from it. Although John's V1 map was spared, the effects of his brain damage were catastrophic for him and his family.

To comprehend the nature and depth of John's troubles, consider the simple drawing in Figure 30. When scientists showed John this picture and asked him to name the object it depicted, he simply stared at it.

"I have not even the glimmerings of an idea," he said. "The bottom point seems solid and the other bits are feathery. It does not seem to be logical unless it is some sort of a brush."

Lest you think that John had never come across a carrot before, here is John's response when he was asked, on a separate occasion, to define what a carrot is. "A carrot is a root vegetable cultivated and eaten as human consumption worldwide. Grown from seed as an annual crop, the carrot produces long thin leaves growing from a root head; this is deep growing and large in comparison with the leaf growth, sometimes gaining a length of 12 inches under a leaf top of similar height when grown in good soil. Carrots may be eaten raw or cooked and can be harvested during any size or state of growth. The general shape of a carrot root is an elongated cone and its color ranges between red and yellow."

John was smart, articulate, and entirely reasonable. When he was asked to identify the object, he could see and describe the drawing before him, and he could logically conclude that an object with one feathery end and one solid end might well be a brush. But he could not simply look at the drawing and recognize what it depicted. Al-

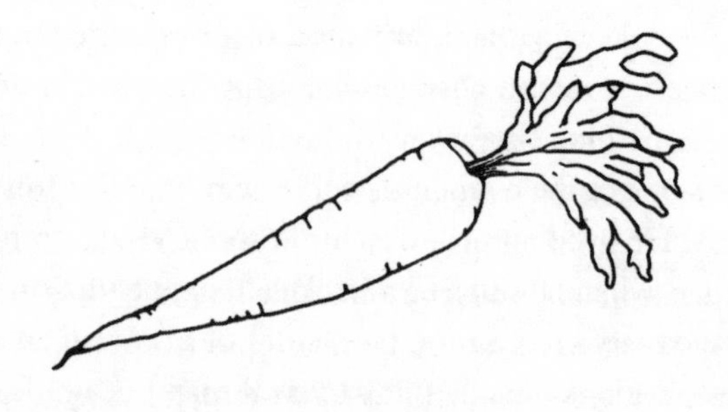

FIGURE 30. The drawing of a vegetable that John was unable to recognize. *Paul Kim, adapted from* Journal of Experimental Psychology, *vol. 6, no. 2. Copyright © 2016 by American Psychological Association.*

though he could see every individual line that made up the drawing, he could not recognize the carrot that they formed. As his predicament illustrates, there is a world of difference between seeing something and knowing what you see.

John's problems went far beyond naming sketches of vegetables. After his stroke, he lost the ability to read. He could not recognize simple objects in his environment, like a razor, a stapler, or a coat hanger. He struggled to recognize places too, including his own home and neighborhood. As John's wife, Iris, put it, "He cannot find his way around our local town or recognize any roads when we drive around the local area which he has known for over twenty years . . . Sometimes in the town he thinks he knows where he is, but unfortunately, he is never right."

Then there was the problem of recognizing faces. He described this as his most embarrassing problem. "I cannot recognize my wife except by sound of her voice, nor my grandchildren, nor family nor friends . . . Awaiting my wife's exit from supermarkets, I have astonished strange ladies by picking up their shopping and walking away with it, under the impression that it was my wife I had been watching pass through the pay desk!" He fared no better with the sight of his own face in the mirror. "Well, I can certainly see a face, with eyes, nose, and mouth et cetera, but somehow it's not familiar; it really could be anybody."

John's substantial everyday challenges did not stem from deficient memory, reasoning, or attention. They came from failures of recognition – the ability to know what or who something is. The word *recognition* comes from the Latin for "to know again." This root meaning captures the essence of recognition: it involves a kind of knowledge that it is built upon the foundation of prior experience. The distinction between sight and recognition is one that most of us rarely have to make. We seem to recognize familiar objects, people, or places instantly and effortlessly, so that it is easy to think of seeing and recognizing as two sides of the same coin. Medical reports describing people like John reveal that they are not the same at all.

Recognition is not limited to knowledge gleaned through sight.

Humans can recognize people, places, or things by seeing, hearing, touching, tasting, or smelling them. But people, or at least sighted people, tend to rely more on vision than the other senses when recognizing the people and stuff around them. For cases like John's, in which this ability fails, a patient is said to have visual agnosia.

John's stroke wiped out large sections of his brain in both the right and left hemispheres. Other patients have been more fortunate and suffered far less extensive damage. As you might expect, less extensive brain damage is generally accompanied by fewer or less severe impairments. But over time, psychologists have noticed a pattern in the specific ways that brain damage impairs recognition. In the case of folks like John, there might be complete visual agnosia, leaving them unable to recognize virtually everything around them. But often, patients have visual agnosia only for *certain types of things.* The most dramatic distinction has to do with whether the things to be recognized are *living things that move of their own accord* or *nonliving inert things.* For example, one person might fall flat at identifying pigs, snakes, and whales but have no trouble naming a chair or a water jug. Another would show the opposite pattern. When the brain damage is quite circumscribed, the impairments may be limited to more specific areas: a patient might struggle only to identify other people or only to recognize printed letters and words. Keep in mind that all these patients had normal recognition abilities before brain damage occurred. They could recognize their child's face or read a handwritten note before injury stole the ability away.

The fact that brain damage can wipe out the capacity to recognize certain categories of stuff but not others tells us that different bits of brain support recognition of different kinds of things. It also tells us something about how the brain divvies up the kinds of stuff that populate our world. Consider the creatures and objects around you right now. There are virtually infinite ways you might group them into categories: *Things that make me feel happy. Things that are speckled. Things with names that begin with a vowel. Things that bit me on my thirteenth birthday.* But those are not the divisions that your brain draws. Above all else, your brain categorizes stuff based on whether or not it is animate, or can move of its own accord. Beyond this major division, there are subtler distinctions. Among animate objects, is it a kind of animal, a part of the body, or a face? Among inanimate objects, is it a small object like a tool that you can manipulate, or is it a large object like a building? Patients can experience very different practical challenges based on the specific type of thing they are left unable to recognize. For example, someone unable to recognize buildings might become lost on a stroll, whereas someone unable to recognize faces might instead feel lost at social gatherings.

Based on reports of these dramatically different impairments, scientists have long had reason to suspect that all forms of recognition are not equal in the human brain. But it wasn't until the 1990s that they first had the chance to watch recognition in action in the healthy human brain. That became possible with the invention of functional MRI brain scans, which can indicate when and where activity changes in the brain. Several of the pioneers who used functional MRI in this way set out to find the seat of recognition. Early work suggested that showing people pictures of faces while they were being scanned resulted in particularly strong activity near the back and sides of the brain, especially in the right hemisphere.

In 1996, a young scientist named Nancy Kanwisher, then at Harvard University, set out to identify specific brain areas that make face recognition possible. Working with Marvin Chun and Josh McDermott, she used functional MRI to identify regions of the brain that were specifically and consistently active when a person in the MRI machine was looking at pictures of faces. She later described the process of discovery: "An initial scan with me as the subject found a promising blob on the bottom of my right hemisphere . . . the signal [there] was higher during the periods when I was looking at faces than the periods when I was looking at objects. Still, a single result like that could have been a fluke. So Marvin and Josh scanned me again. And again. And again. To our delight, the trusty little blob showed up in exactly the same place every time."

Kanwisher and her colleagues went on to find that trusty little blob in many other people's brains. They would even give it a name —the fusiform face area, or FFA, named after the specific fold of cortex within which it was found. Since then, countless functional MRI studies have observed the FFA in adults and demonstrated that it is nearly universally found in approximately the same area, at the bottom of the temporal lobe in the right hemisphere. Some people also have a similar face zone in the left hemisphere, but it is nearly always smaller than the one on the right.

When brain damage wipes out a person's FFA, they specifically lose the ability to recognize faces. Such a person would be able to tell a carrot from a hairbrush and recognize familiar places in the neighborhood, but they would share John's struggles with recognizing their spouse, their children, or even their own reflection in the mirror. This special type of visual agnosia has a name: prosopagnosia, from the Greek for "lack of face knowledge." People with severe prosopagnosia experience enormous social challenges. They find themselves adrift in a sea of unfamiliar faces even when they are among friends and family. They must rely on cues like hairstyles, facial hair, clothing, or voices to piece together the identities of those around them.

Even among healthy people who have never suffered brain damage, there is an enormous range of ability to recognize faces. An estimated 2 percent of the healthy human population is profoundly unable to recognize faces, despite having intact vision and no evidence of brain damage. Often these people experience social anxiety or embarrassment. Scans show that healthy individuals with prosopagnosia do have an FFA, but their struggles may be linked to subtle structural differences within it. Regardless of slight differences, evidence suggests that these folks aren't suffering from a medical condition; they just happen to be at the farthest end of a broad spectrum of human face-recognizing ability. Some people can never recognize a face, whereas others can never forget one.

Taken together, brain scans and brain damage have given scientists plenty of proof that the FFA is important for face recognition.

Still, there is nothing more convincing than observing how moment-to-moment perception can be altered by tweaking the activity in a person's FFA. That's what happened to a forty-five-year-old man I will call Terrance, who was preparing to undergo brain surgery. Like many of the patients whose stories appear in this book, Terrance suffered from powerful seizures that could not be controlled with medication. His surgeon had implanted electrodes in his brain to identify where the seizures began, with the hope that the team could find the source and remove it. But it just so happened that two of the implanted electrodes rested in Terrance's FFA. The medical and scientific team wanted to know what would happen if they sent a tiny jolt of electricity into those electrodes, stimulating this part of the brain. Would Terrance see or experience faces differently?

The neurosurgeon, Josef Parvizi of Stanford University, asked Terrance to look directly at his face. Seated in his hospital bed, Terrance appeared tired and almost grim. His broad shoulders were draped with a hospital gown, and his head was wrapped in layers of gauze that covered the bundle of wires leading to the electrodes.

There was a loud click, and the electric jolts hit their mark. Terrance's face suddenly brightened. The corners of his mouth curled up to form a subtle smile as he shook his head in astonishment.

"You just turned into somebody else," Terrance said. "Your face metamorphosed." He gestured to illustrate on his own face. "Your nose got saggy, went to the left. You almost looked like . . . somebody I'd seen before but somebody different." He raised his eyebrows and nodded his head. "That was a trip."

"Hmm," said the surgeon. "Could you see my eyes in place?"

"I could see your eyes, but you could have been somebody else who had, well, he had similar eyes to Dr. Parvizi, but you were someone else." He shook his head again and lifted his hands to his cheeks. "Your whole face just sort of metamorphosed."

When the team was ready to try again, Parvizi counted down with Terrance. "All right, ready?"

"'Kay," Terrance said quietly.

"One . . . two . . . three," said the surgeon. There was another click.

Terrance drew in a deep breath, nodded his head, and smiled a puzzled smile. "Yeah. It metamorphosed again, and you looked like someone I've seen before but maybe a different person in my memory. Almost like your nose kind of shifted to the left a little bit, and your look just changed."

"Did I keep my gender?"

"Yeah, oh yeah."

"How did you know I'm not a female?"

"Because you're still wearing a suit and tie."

"Oh, you could see the suit and tie?" the surgeon said with a little chuckle.

Terrance nodded and grinned. "Yeah. Only your face changed. Everything else was the same."

"Did the position of my lips and nose and eyes stay the same when they got warped?"

"They shifted, let's say they shifted to a side and maybe stretched, but they didn't get larger or smaller." He cupped his hands and held them out, as if he were trying to take hold of a description and hand it to the physician. "It was more of a *perception,* how I perceived your face."

"Interesting," the surgeon said. "Tell me more."

Terrance smiled and gave a quick shake of his head. "That's about all I can say. All of a sudden you were you and then you weren't you. You could have been someone else who was standing there in front of me."

Terrance's experience must have truly been astonishing – "a trip," as he described it. But it also proved an important link between activity in the FFA and recognition of facial identity. A great deal of work, both with patients and healthy participants, suggests that the FFA is essential to processing the relative positions, or configurations, of parts within a face. If that sounds like a strange specialty, consider how difficult facial recognition really ought to be. Virtually every face you have ever seen has the same basic set of parts in the same basic arrangement. The only reason that they don't all look virtually identical to you is because you have help from specialized areas of the brain, like your FFA.

The FFA is not the only area that specializes in processing faces. For example, another area in the occipital cortex specializes in identifying the parts of the face, rather than their relative spacing, and may help more with detecting *whether* something is a face than identifying *which* face it is. Other face-specific regions have been found as well, including one that specializes in identifying the emotions expressed by a face. These regions each play an important role in your ability to extract crucial information from a face. Collectively, they allow us to meet the challenges of recognizing and reading faces with ease.

Yet faces were just the beginning. Using functional MRI, scientists also began identifying areas of the brain that specialize in processing other important types of things. Kanwisher and her colleagues were involved in many of these discoveries. They helped characterize an area that gives a strong response when participants view houses, buildings, and indoor or outdoor scenes: images that offer clues about a particular environment. They named this zone the parahippocampal place area, or PPA, after the fold of tissue within which it was found. When brain damage destroys this region, patients may be left unable to recognize landmarks in their environment.

Scientists have also learned about the PPA by observing what happens when it is electrically stimulated. A medical team did this in one young patient who, like Terrance, was preparing to undergo surgery for seizures. When the stimulation was applied, the surgeon asked if the patient felt or saw anything. The patient looked confused and put his hand to his forehead. "Yeah, I feel like I . . . I feel like I saw, like some other site, we were at the train station." Just as stimulating Terrance's FFA changed his perception of a face, stimulating this young man's PPA changed his perception of his environment. And just as scientists have pinpointed several distinct regions that support face processing, researchers have now found a handful of brain areas, including the PPA, that contribute to processing information in a visual scene.

A third major revelation came when Kanwisher and colleagues identified a brain area that responded specifically to pictures of bodies or particular body parts, such as an arm or a foot. They named this brain area the extrastriate body area. Several studies since then have investigated this area by means of transcranial magnetic stimulation, a technique that applies strong magnetic pulses at the scalp to temporarily disrupt brain activity on the surface of the brain beneath. In general, disrupting this area interferes with people's judgments about a body's shape and configuration.

Just as we tend to underestimate the complexity involved in recognizing a face, we are largely unaware of the complexity of a body's configuration and movement and the challenges of interpreting them. Bodies, like faces, contain a wealth of social information. Body shape and movement can tell us *who* a person is, even if that person is looking the other way or is far off in the distance. Posture, configuration, and movement can tell us about a person's goals, next actions, and even emotional state. You "read" people's bodies far more than you might expect, in order to understand what they want and feel. As such, it is hardly surprising that our brains have specialized areas, like the extrastriate body area, to aid us.

In my own research with Dr. Kanwisher, we identified and named a second region, the fusiform body area, which also contributes to this process. This area is found immediately next door to the FFA in the tissues at the bottom of the temporal lobe. I became fascinated by these twin areas, the fusiform face and body areas, always next to each other in the brain just like faces and bodies are next to each other in everyday life. It sparked my interest in how and where these zones developed in the brain. And that interest, in turn, is what led me to the topics and questions explored in this book.

Since the discovery of the extrastriate and fusiform body areas, scientists have identified people who have suffered damage to one or both of these areas. These patients can help us understand what the body areas do by revealing how people are affected by their destruction. The scientists found that these patients had problems remembering and distinguishing between pictures of different body parts.

The list of object categories blessed with their own zones in the object map does not end with faces, scenes, and bodies. For example, scientists have discovered regions that specialize in representing handheld tools. Others have identified a zone involved in processing the shapes of written letters and words, but only in the brains of literate people and only for letters from the alphabet that they learned to read. There are also nearby regions of the brain that play a more general role in processing object shapes and recognizing objects that do not belong to one of these categories.

This hodgepodge of object types and their regions may seem overwhelming. What are we to make of these strange zones for different kinds of stuff? To answer this question, it helps to consider the phenomena and dimensions that underlie recognition. Whereas the brain uses maps to represent continuous phenomena like location in space or sound frequency, it tends to represent categorical phenomena like molecular identity by using specialized zones, distributed codes, or some combination of the two.

Recall that stuff can be grouped in many ways. You can judge how similar two things are by using any number of possible dimensions. The particular dimensions that you choose determine how you categorize the stuff. The same is true for regions of the brain. Two areas of the catfish brain provide a now-familiar example; one contains zones for odors based on molecular structure, whereas the other area groups odors by behavioral relevance (that is, whether the odor signals the presence of food or fellow fishes). Despite superficial differences, the recognition of visual objects is a lot like tasting or smelling. In all cases, the goal is to figure out what kind of stuff you are faced with. And in all cases, that stuff can be grouped according to a variety of dimensions. The obvious question, then, is which dimensions matter?

Even before considering the evidence from brain scans, you can take a first pass at answering this question based on your intuition as to how objects might or might not be grouped or combined. Imagine, if you will, an object that is half house and half lip balm. Can you do it? What did you come up with? A tube of lip balm spackled to the side of a house? A house made out of petroleum? Or maybe lip balm in a case that is shaped like a miniature house? All these ideas fall short of meaningfully integrating these two types of objects. That does not mean that all objects are incompatible. There *can* be a hand-held object that is both spoon and fork or an item of clothing that is both skirt and shorts. (Yes, I'm thinking of you, spork and skort.) There can also be animals that are both leopard and tiger or donkey and horse (liger and mule, respectively), not to mention imaginary creatures like the griffin or the centaur. And if you have ever played in a tree house or slept in the cabin of a train car, you will know that houses can be combined with other sorts of things as well.

Given that many types of objects can be integrated, what stands in the way of making or even imagining a lip-balm house? To begin with, houses are big. We can use them as landmarks to find our way around town or as shelters to enter and inhabit. We never have and never will pick them up. Unlike a tube of lip balm, a house cannot be slipped into our pocket. We do not associate a taste or a texture or a part of our body with houses. There is no familiar way to grip a house, to twist it, or to spread it. In short, the wholly different ways that we use and interact with these two objects render them incompatible. We can distill all the individual differences between them to a single one: scale. We move about relative to large things, whereas small things can be picked up and moved about relative to us. One sort of thing is useful for telling you where you are and the other for enabling you to take, hold, and use it.

But there is a third kind of stuff in our world: unpredictable stuff. By that I mean living, animate things: flies, spiders, hamsters, cats, cows, and our fellow human beings. Often we cannot simply take hold of other creatures and do as we wish with them. At best, they may resist or surprise us. At worst, they may bite, trample, or ostracize us. A garden snake might resemble a garden hose in shape and diameter, but you approach and reach for the two very differently. In short, we can't expect to just act upon creatures; we must also be prepared to react to them. This is true when we interact with animals, but perhaps even truer when we interact with other humans.

To safely navigate social interactions, we need to continuously collect and assess social information. We need to know who a social partner is, so that we can determine our shared history, recollect the person's prior actions, and predict how he or she may act today. We need to cull information from a furrowed brow, dilated pupils, or an expansive body stance to infer the person's mood and intentions so that we may improve our predictions for what they'll do next. It would be absurd to seek such information from a building or a personal hygiene product. There is a clear and fundamental divide between the set of things we act upon and the set of things that interact with us.

This intuitive distinction between types of objects provides a helpful framework for understanding the bevy of object recognition zones scientists have found in the human brain. Altogether, these zones cover a giant swath of cortex at the back of the brain, stretching from the inner bottom surface of the temporal lobe, around to the bottom outer edge of the brain, and up the side of the occipital and parietal cortex. Intriguing evidence suggests that the zones are islands within a larger, overarching object map. Using functional MRI scans, scientists have shown that this overarching map is organized according to two object dimensions: object size and object animacy. This results in three major divisions in the object map: large inanimate objects, small inanimate objects, and animate creatures of any size. The map is laid out symmetrically, with large objects represented at the far ends of the elongated map and with separate representations for small objects and creatures of all sizes nestled in between them, as illustrated in Figure 31. The specific zones that specialize in processing faces and bodies, such as the FFA, lie within regions of the overarching map that represent living things. Zones, including the PPA, that prefer buildings and other landmarks lie in portions of the map that prefer large, inanimate objects.

This object map offers us a brainscape of the people and stuff in our world and—most important—their relevance to us. The most relevant stuff, which requires adept recognition, is represented within large specialized zones. This is like the magnification of hands in the human S1 touch map or the magnification of foveal inputs in the V1

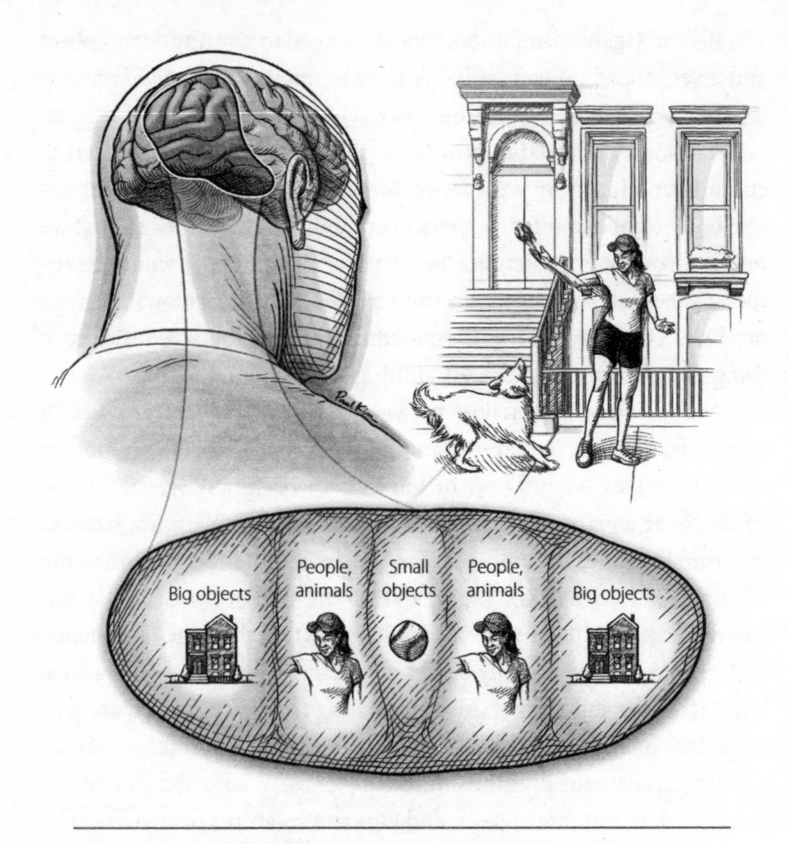

FIGURE 31. An illustration of the major zones within the human object map. *Paul Kim*

visual map. Crucial stuff gets extra space. And for humans, other people are crucial. Recognizing the identity of the faces around us, and the expressions on them, is of paramount importance, not only if we hope to be popular, but if we hope to stay alive. So too is recognizing the environment around us: where we are and how we might get elsewhere. But perhaps the most important thing we can take away from the object map is an appreciation of how we are shaped by our world. The natural properties of the objects and creatures around us offer us specific opportunities for acting upon or interacting with them. In turn, these opportunities, and our experience of seizing upon them, shape the layout of this special map within the brain.

THE MAKING OF OBJECT MAPS

Where do the specialized zones in your object map come from? Given that the fetal brain contains visual and auditory maps even before seeing and hearing begin in earnest, could babies be born with zones for objects that they have not yet seen? This question has been hotly debated for quite some time. Finally, we now have tantalizing clues as to how zones within the object map do (and in some cases, do not) develop.

Turn back the clock to your time in the womb. The retinal waves that trained your early visual maps did not stop when they reached V1. But by the time these waves had journeyed from your retina through your thalamus and V1, out into the suburbs and then the exurbs of your V1 visual map, the strength and cohesion of these waves had lessened a great deal. The effect of these weak waves on the outlying territories of your immature visual cortex was to give them a slight preference for inputs from a particular region of the visual field. Some of these cells also had a slight preference for movement or basic visual features like shape or line curvature. But by and large, these cells were chemically and structurally immature, weeks behind those in your V1 visual map, your A1 sound-frequency map, and your other primary sensory maps in developing strong preferences and producing powerful signals. At the time of your birth, your nascent object map was something of a no-man's-land of fledgling neurons awaiting guidance.

Once you entered the world, you had visual experiences upon which to train this no-man's-land of cells. Newborns don't have that much control over what they look at; a small baby generally sees what adults put directly in front of its face. And for the most part, that is the adult's own face. Babies younger than three months of age overwhelmingly see close frontal views of their caregivers. One study that recorded infants' visual environments logged fifteen minutes of close views of faces for every hour of footage. The position of human nipples might also factor into early face exposure, as it places newborns in close proximity to their mother's face during prolonged periods of nursing. Such an up-close and intense early exposure to faces is the perfect way to teach that expanse of immature visual neurons, particularly those that already slightly prefer foveal inputs, to dedicate a hefty chunk of real estate to representing and processing faces. Crucially, this real estate will include the future home of the FFA face zone.

To better understand how specialized zones develop within the object map, a team of neuroscientists studied infant macaque monkeys. Adult macaques have an object map similar to that of adult humans; both include special zones involved in processing images of faces, body parts, objects, and scenes. The research team presented pictures to newborns macaques while scanning their nascent object maps with functional MRI. The scans revealed no trace of specialized zones for faces or objects. There were distinct patterns of neural activity for the retinal location of images presented to the newborns, but not for the types of objects they were shown. Yet this no-man's-land would be settled soon enough. Specific and stable responses to faces and objects emerged gradually in the five to six months after an animal's birth.

Does this dramatic specialization in the first few months mean that monkeys need to be exposed to faces during this window in order to develop zones like the FFA? To find out, scientists raised three monkeys from birth so that they did not see either human or monkey faces. The humans who cared for these monkeys wore welder's masks to obscure their faces. When the monkeys were about three months old, they glimpsed a handful of pictures of people with faces — their first time ever seeing a face. You might imagine they would stare at the faces, since faces were new to them. Instead, the monkeys looked at the *hands* of the people in the pictures more than their faces — the opposite of what normal monkeys at their age would typically do. Five months later, functional MRI scans revealed that these monkeys had no zones dedicated to faces. Without exposure to faces in infancy, these monkeys did not develop such zones in their object maps. However, they did develop zones for the most dynamic, engaging, and socially relevant thing in their environment: body parts, specifically hands.

Although these studies involved monkeys, evidence suggests that the same process is at work in humans. Functional MRI scans of human infants have shown that the basic regions devoted to processing faces and scenes can be detected in babies four to six months old. Throughout infancy and childhood, responses to different types of objects in the nascent object map become more stable and distinct. Specialized zones representing privileged categories like faces or bodies expand, taking up larger portions of the overall map. Face areas in particular continue to grow larger and more specialized throughout adolescence, in parallel with continuing improvements in the ability to recognize faces.

This prolonged period of improvement and neural specialization presumably depends on children's exposure to faces, although no experiment has proved it. Scientists can't ethically prevent a child from seeing faces during the early months or years of life. But some children are born or raised in circumstances that naturally alter their exposure to faces. Babies who are born with cataracts blocking their vision fall into this category. Their cataracts may be surgically removed, but the timing of the surgery impacts how their brain maps develop. In particular, babies whose left eye is temporarily blocked by a dense cataract may never develop the ability to recognize faces as well as they otherwise might. By depriving the left eye of glimpses of faces, the cataract deprived the right hemisphere of face information during those earliest months of life. Even after surgery and plenty of exposure to faces in subsequent years, these individuals tend to show at least subtle impairments at face recognition. These continuing challenges suggest that there is something special and irreplaceable about exposure to faces in the first few months of life.

The problem may be that, after four months of age, the brain is not as receptive to developing and refining zones for face processing. Or it might be that, after that time, the baby has lost its chance to live in a face-filled world. Although the first four months of a baby's life are dominated by views of people's faces, the situation changes rather quickly as the baby learns to sit up, crawl, reach, and manipulate objects. Views of faces dramatically decline. In their place, children receive far more views of hands – both their own hands and those of a caregiver, usually in contact with a toy, a cup, or some other object. Children are now exploring, choosing what they interact with, and having visual experiences shaped by their new postures and movements. As views of faces are replaced with views of hands and objects, these new visual experiences may specifically hone object map representations of body parts and small objects more than those of faces.

The development of the object map hits another major milestone when children reach about five years of age, when they begin learning to read. In the process of becoming literate, children develop a new zone that specializes in processing written words and strings of letters specifically for the alphabet and language they are learning to read. This zone develops immediately next to the FFA in the left hemisphere and, with time, steals away some of its face territory, converting neurons that once preferred faces into ones that prefer written symbols. At the same time, the right FFA grows larger in these children, as if to compensate for the loss on the left.

When people learn to read in adulthood rather than childhood this reconfiguration of face zones in the object map does not occur. By this time, maturation and experience have probably crystallized the face zones, preventing any possible theft from the left FFA. Of course, this does not mean that people cannot learn to read in adulthood; many do. But it helps explain why the process tends to be more difficult and to result in slower, less fluent reading ability. It also demonstrates how the principles at work in maps like M1 and A1 come into play in the object map; territories within the map are claimed or relinquished in a cutthroat competition to capture brain space. When the clock runs out and the dust settles, the allocation of terrain to processing *something* – be it a certain type of object or a certain range of sound frequencies – has profound effects on our ability to detect, discriminate, and identify.

This cutthroat process of allocating terrain in development does not mean that adults can't learn about new types of objects. Although islands within the mature object map are devoted to processing certain special types of stuff like faces, that still leaves space for representing less privileged types of objects or for learning about new objects, even in adulthood. The representation of new objects is carried out using a distributed code, like the code used in the piriform cortex to represent odors. From day to day, you may encounter new gadgets and, through experience, introduce them to your object maps. Distributed codes give you the flexibility you need to recognize new kinds of stuff.

One clever experiment showed how neural representations of new objects became less distributed as people learned more about the objects and their uses. Specifically, scientists trained adults to use contraptions built from children's construction toys as tools to poke, push, lift, or scatter other things. When the adults saw pictures of the contraptions before their training, functional MRI brain scans revealed a scattered or distributed pattern of activity across the participants' object maps. But when they were scanned again after the training, viewing pictures of the contraptions specifically sparked activity in the island of the object map that specializes in tools. In other words, viewing an identical picture triggered different activity in their object maps before and after they learned about what the makeshift gadget in the picture does. Their representation of the gadget changed not because the visual information changed, but because their knowledge and experience did. This result beautifully illustrates how recognition goes far beyond visual perception.

In fact, scientists have begun to wonder if it makes sense to think about the object map and its constituent zones as visual areas at all. They are by no means limited to processing inputs from vision. If you close your eyes and hear the spoken word "hammer" or hear the sound of a hammer striking a surface, activity will ramp up in your tool zone. The same is true if you pantomime the hand motion of using a hammer or simply imagine using one. Moreover, people who have been blind since birth have object maps just like sighted people do. Many aspects of the organization and location of their object maps are quite similar to those of sighted people, even though one set of people relies heavily on vision to recognize objects and the other set has never seen an object.

Although a great deal remains unknown about how the object map and its dedicated zones carry out the important work of recognition, the identity and organization of these specialized zones offer fascinating clues. The major distinctions of size and animacy in the map reveal a truth that has been hiding in plain sight: the properties inherent to each object determine how we can interact with it. Everything in our world offers precious information, but exactly what information we extract from it will depend jointly on the characteristics of the object and our own needs. The islands that we all develop for recognizing faces, bodies, and scenes reveal how vital social and environmental information is to us, each and every day. We do not come by this information easily; it takes years of learning and experience. And plenty of people among us, for a variety of reasons, do not come by this information at all. The next time an acquaintance fails to recognize you or a friend gets lost on a straightforward route, keep in mind how hard such things by all rights ought to be. It is a wonder of learning and representation every time you manage to know again.

9

Imagining, Remembering, and Paying Attention with Brain Maps

THOUGH OUR VARIOUS SENSES and manifold movements are remarkably diverse, they all reflect concrete physical phenomena. If perception and action are products of the here and now, it would be reasonable to conclude that sensory and movement maps are limited to representing the moment-to-moment events happening around and within us. And yet they are not. Quite the opposite, in fact—brain maps can transport us beyond the here and now, representing things that happened long ago, two minutes ago, or may never happen at all.

The most dramatic example I've seen of the brain's transportive power comes from the description of a young woman on what may have been the scariest day of her life. Wilder Penfield referred to her as MM, but I will call her Miriam. She was awake and alert while a surgical team prepared her for surgery. Her scalp and skull were opened up and her brain exposed, so that Penfield and his colleagues could search for the source of her debilitating seizures. Unlike Penfield's patients whom we have already met, Miriam had seizures that did not begin with movements. Rather, she would be overcome by a "feeling—as though I had lived through this all before." At other times, she would experience flashbacks to previous moments in her life, sometimes moments that she could no longer otherwise recall. After these sensations, she would walk about in confusion or speak unrelated strings of words—actions that she could not remember performing once the seizure was over.

In search of the origins of her seizures, Penfield probed the right side of her brain with an electrode. He began stimulating regions of the temporal lobe, inching inward toward the hippocampus, a structure that lies beneath the cortex and is profoundly important for memory. After one stimulation, she said, "I think I heard a mother calling her little boy somewhere. It seemed to be something that happened long ago." Asked to explain what she heard, she added, "It was somebody in the neighborhood where I live."

When Penfield stimulated the same place again, she said, "I hear the same familiar sounds. It seems to be a woman calling, the same lady. That was not in the neighborhood. It seemed to be at the lumberyard." And then she added, "I've never been around the lumberyard much."

Penfield pressed on, probing new spots in her brain with the electrode. When he stimulated another place, she said, "I hear voices. It is late at night, around the carnival somewhere—some sort of traveling circus. I just saw lots of big wagons that they use to haul animals in."

At another site: "Oh, I had the same very, very familiar memory, in an office somewhere. I could see the desks. I was there and someone was calling to me, a man leaning on a desk with a pencil in his hand."

Penfield began narrowing in on the site where the seizures were originating. Working slowly but confidently with his scalpel, he cut out a large chunk of cortex from the side of the brain and then tested the deeper tissue that now lay exposed. With one stimulation, she said, "I feel very close to an attack. I think I am going to have one—a familiar memory." With another: "Oh, it hurts and that feeling of familiarity—a familiar memory—the place where I hang my coat up, where I go to work."

That is the last we hear from Miriam in her own words. Penfield was finished using her responses and her evoked experiences to locate the damaged tissue where her seizures began. He found it close to the hippocampus. The tissue was hardened, perhaps compressed due to a complication that happened when Miriam was born. Although Penfield did not say, presumably his scalpel went deeper still to cut out the rest of this damaged tissue. We can only hope that Miriam went on to survive the surgery, recover, and return to a better life.

Aside from the dramatic circumstances of Miriam's surgery and the notion that a piece of her brain was removed in the process, something else about Penfield's account is unsettling. Miriam's first-person recounting of her peripatetic mental experiences seems strange and almost magical. With a tiny jolt to her brain, she was transported to different times and places. Her senses informed her of things that weren't there. Perched in the sterile operating room, surrounded by drapes, gauze, and medical staff, she *heard* the voice of someone's mother, *saw* wagons and strange men, and *felt* that she was somewhere else. The whole thing sounds like science fiction.

But though Miriam's circumstances were extraordinary, her experiences beyond the here and now were not. There are many ways in which we are transported, or willfully transport ourselves, to other places and times every day. In fact, we spend about half our waking hours thinking about something other than what we are doing or perceiving at the moment. We regularly hear, see, and feel things that aren't before us. We use our imagination to conjure fictional events or possible futures. We call past sensations to mind, remembering the expression on someone's face or the sound of their voice. We focus on recent sensations or actions, as when we mentally repeat a shopping list or mentally retrace our steps to find a lost set of keys. At night, we succumb to dreams filled with sensations and actions, creatures and emotions—as all the while we drool on a pillow in the dark. Why should we have the capacity to experience things that are not happening right now, at this very moment? And how do our brains carry out this impressive feat? As you might have guessed, the answer lies, at least in part, with the brain's cornucopia of maps.

BEYOND THE HERE AND NOW

On its face, there would seem to be nothing in this world less amenable to scientific study than mental imagery. It is unobservable and immaterial, culled in an instant from nothing, only to blink just as quickly back out of existence. What *is* imagery? Who has it? And how does it work? These are difficult questions to answer. How do you study something that cannot be objectively seen, felt, or measured? And why would scientists be brazen or foolish enough to even try?

One of the first bold attempts to pin down mental imagery and subject it to scientific scrutiny took place in England in the 1870s. People were asked to fill out a questionnaire with unusual mental exercises like this one:

> Think of some definite object—suppose it is your breakfast-table as you sat down to it this morning—and consider carefully the picture that rises before your mind's eye.
>
> *Illumination.*—Is the image dim or fairly clear? Is its brightness comparable to that of the actual scene?
>
> *Definition.*—Are all the objects pretty well defined at the same time, or is the place of sharpest definition at any one moment more contracted than it is in a real scene?
>
> *Colouring.*—Are the colours of the china, of the toast, bread crust, mustard, meat, parsley, or whatever may have been on the table, quite distinct and natural?

Although your morning meal may not have involved china or bread crust, presumably you can conjure a memory of where you were and what was around you earlier today. However you experience this memory—how you may see it in your mind's eye—will seem normal to you. But does everyone remember and visualize past places and events in the same way? Before this quirky questionnaire was developed, no one had bothered to ask. Other exercises in the survey, like the one that follows, took on other forms of mental imagery to explore how its respondents experienced them through different senses.

Call up before your imagination the objects specified in the six following paragraphs, numbered A to F, and consider carefully whether your mental representation of them generally, is in each group very faint, faint, fair, good, or vivid and comparable to the actual sensation:—

A. Light and colour.—An evenly clouded sky (omitting all landscape), first bright, then gloomy. A thick surrounding haze, first white, then successively blue, yellow, green, and red.

B. Sound.—The beat of rain against the window panes, the crack of a whip, a church bell, the hum of bees, the whistle of a railway, the clinking of tea-spoons and saucers, the slam of a door.

C. Smells.—Tar, roses, an oil-lamp blown out, hay, violets, a fur coat, gas, tobacco.

D. Tastes.—Salt, sugar, lemon juice, raisins, chocolate, currant jelly.

E. Touch.—Velvet, silk, soap, gum, sand, dough, a crisp dead leaf, the prick of a pin.

F. Other sensations.—Heat, hunger, cold, thirst, fatigue, fever, drowsiness, a bad cold.

The entire questionnaire, from the mustard and parsley to the crisp dead leaves, was the brainchild of Francis Galton, an expansive thinker with his fingers in many pies. He studied and wrote about statistics, meteorology, human perception, and psychology. Today he may be best known as a founder of eugenics, a pseudoscientific and political movement that declared some races genetically and mentally superior to others and advocated societal interventions to promote "superior races" and their "superior genes" at the expense of everyone else. This horrifying movement was used to justify racial subjugation, forced sterilization, and even genocide.

Galton himself coined the term *eugenics* in one of his books. Later, in the very same book, Galton devoted an entire chapter to revealing the results of his groundbreaking survey on mental imagery. The questionnaire had been circulated to hundreds of people, including scholars, artists, schoolchildren, and passersby on the street, but Galton paid special attention to a sample of responses from a hundred adult men, "at least half of whom are distinguished in science or in other fields of intellectual work." To Galton, this sample represented the best that humankind had to offer: aristocratic European men.

If Galton had hoped to find that these ideal subjects all experienced either strong or weak mental imagery, demonstrating which one was the hallmark of superior intellect, he was sorely disappointed. Their ability to conjure mental images ranged the full gamut. For example, one reported, "All the objects in my mental picture are as bright as the actual scene." Another described imagery that was "fairly clear, but not equal to the scene." A third replied, "My powers are zero. To my consciousness there is almost no association of memory with objective visual impressions. I recollect the breakfast-table, but do not see it." The most apparent generalization to be made from Galton's sample was that very little could be generalized. Individuals differed substantially in how vivid and clear their mental imagery was, if they were able to generate such a thing at all.

Galton's examination of mental imagery was groundbreaking for its time, reporting the first study to offer an in-depth assessment of people's abilities to generate imagery. It also described the pioneering use of a format that has now become a hallmark of modern research: the questionnaire. But like many of the early scientists in psychology and neuroscience, Galton sought to open windows onto the human mind and brain while closing the door on most of humanity. His work reminds us that the pursuit of scientific truth is always embedded within a context—a time, a place, and a tangle of interlocking beliefs and biases. The challenge is looking with a ruthless eye through what the past has left us, to sift and sort: to both recognize where there is value and to clearly see and call out where hatred and fear parade about under the mantle of science.

From Galton's survey, and considerable work that has come since, we know that people do not all experience mental imagery in the same way and to the same degree. But most people do experience mental imagery in some form and to some degree. Some people even experience such vivid imagery that they find it distracting or disruptive. In all cases, something sparks these imagined sensations: it might be a wishful daydream, following Galton's written exercises, or even receiving an electrical jolt directly to the brain, as was the case for Miriam. Imagery can be triggered in many ways. But what is it made of, and why does it feel so much like perception?

The answer to this question is surprisingly simple. Mental imagery feels like perception because it engages many of the areas of the brain involved in actual perception, and it engages them in much the same way. Modern technologies like functional MRI brain scans have led to breakthroughs in our understanding of how mental imagery works. These scans revealed that activity within the V1 visual map itself reflects the contents of a person's mental imagery, thereby literally making a picture of the imagined content in the person's brain. The illustration in Figure 32 shows how imagery and perception generate similar patterns of activity within the V1 visual map, although the activity is weaker for imagining than it is for actually seeing.

V1 is by no means the only visual area of the brain involved in making mental imagery. For example, V1's neighbor, a visual map nicknamed V2, is also in on the game. Imagining visual motion also ramps up activity in a motion-preferring zone of the visual cortex. Imagining a face boosts activity in the FFA face zone, while visualizing a place activates the PPA place zone. Both within and across brain maps, imagining something involves weakly activating the same bits of brain required for you to see it.

Consider that for a moment. Vision is triggered from patterns of light entering your eye and stimulating your retina. Mental imagery is generated entirely from *within*, by a dark stew of brain cells that have never been touched by so much as a single ray of light. And yet that dark stew can spark nearly the same type of activity in your visual cortex as your eyeballs can. In V1, that activity forms a literal picture of what you imagined, although drawn with firing frequencies rather than ink. As far as your visual cortex is concerned, imagining resembles a weaker form of actual seeing.

And that is just the beginning. You can use your imagination to generate imagery for senses other than vision as well. Whatever type of imagery you generate, you do so by activating the appropriate brain maps, mimicking the specific activity that would arise for actually perceiving the imagined thing. When you imagine a sound or hear a song in your head, this imagery is evident in the activity within sound-frequency maps, including A1. When you imagine being touched on a part of your body, touch maps, including S1, allow you to feel that imagined sensation. When you imagine speaking, you are generating activity in areas that contribute to actual speech production. When you imagine moving your fingers or your hand,

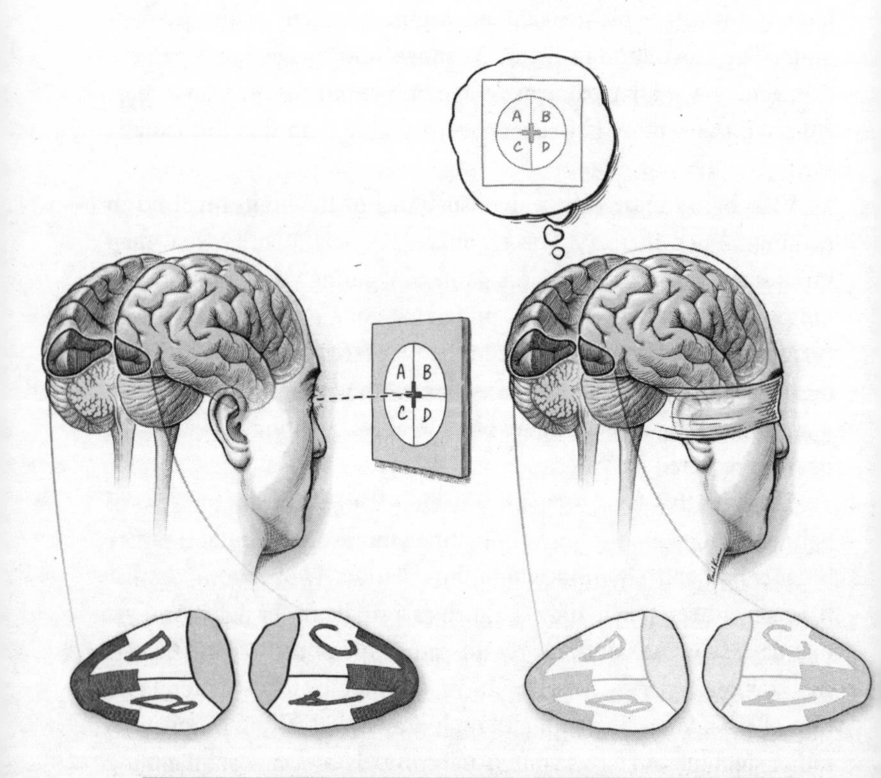

FIGURE 32. A comparison of visual perception and mental imagery in the V1 visual map. *Paul Kim*

neurons in sectors of your motor cortex, including the hand region of M1, ramp up their firing rates.

This neural mimicry is extraordinary in its own right, but it is all the more remarkable when you consider how often it is happening in your brain. For most of us, mental imagery dominates the time we spend immersed in thoughts beyond the here and now. Daydreaming is one obvious way that mental imagery finds its way into our daily lives. Daydreams can help us consider and prepare for possible future outcomes. But imagery is not just the stuff of daydreams. For example, people often experience visual, auditory, or other kinds of imagery while reading or listening to narrative passages or stories. This imagery is accompanied by activity in brain maps for vision, hearing, and even touch. By way of this imagery, you can virtually see, feel, and hear the world that the storyteller or author has created for you.

In the same vein, when most people retrieve memories, they experience imagery of past people, places, and events. Activity in your visual, auditory, and other brain maps represents this recollected imagery just as it would for imagery manufactured out of whole cloth from your imagination. If we weren't so accustomed to using mental imagery to access information about the past, we might appreciate how remarkable a feat this is. You may not be able to visit your childhood bedroom in person; perhaps it has even been razed to the ground. Nonetheless, most people have the ability to feel as if they are in that room again—to see its layout and perhaps notice details in the objects around them. As far as your mind and your brain are concerned, you are practically there, even if the place no longer exists outside your mind.

The immersive nature of memory can be wonderful or terrible, depending on whether the memories are ones that you care to revisit. For example, people suffering from posttraumatic stress disorder or depression often experience intrusive and upsetting imagery of past events. These painful or frightening recollections would not have such power if memory were a simple recounting of *what* happened and not how it happened—and crucially, how it *felt* to have it happen to you.

Mental imagery allows us, for better or worse, to reach back into our distant past. But we also harness it to keep track of things that have only just happened. Psychologists call this working memory—the ability to keep information "on hand" for a short period of time. Let's say you see a speeding car crash into a parked car and then careen away. You are the only witness. You glimpse the license plate number: NJ612B5. You need to remember the number until you can write it down or report it. What do you do? If you don't have a pen or a cell phone handy, you might repeat the string of numerals in your head: *NJ612B5. NJ612B5. NJ612B5.* This inner speech is a kind of mental imagery.

Psychologists have studied this phenomenon for the better part of a century and have made several intriguing observations. It is not the meaning nor the written appearance of the words or numbers you hope to remember that matters for this type of working memory, but rather how the words or numbers sound when spoken aloud. Trying to remember a string of words that sound similar, like *bite blight trite try tine* is much harder than remembering words in which each sound is different. Likewise, it's the length of the word when *spoken,* rather than its length on the written page, that matters for this kind of working memory. People can only mentally rehearse content that takes less than about two seconds in total to say. If it runs longer than that, they forget the beginning by the time they reach the end.

Studies using functional MRI brain scans have revealed the role of brain maps in working memory and, in doing so, helped to explain its peculiarities. These studies showed that you use body maps in your motor cortex and sound-frequency maps in your auditory cortex to imagine speaking and hearing the content, respectively. That's why actually talking or singing interferes with your ability to mentally rehearse; real talking and imagined talking use many of the same bits of brain. Likewise, listening to someone else actually speaking interferes with your ability to represent the sounds of your own inner speech using the auditory cortex. In short, these brain areas can process speech movements or sounds that are either real or imagined, but they fail dreadfully at doing both at once.

We see a similar process at work in a second form of working memory that you might use to maintain visual information, like the pattern on a fabric or the layout of pieces on a chessboard. Looking at an image, imagining that image, and maintaining that image in working memory all evoke activity in the same bits of brain within your visual maps. And just as working memory based on inner speech is disrupted when you hear actual speech, your ability to hold an image in mind is disrupted when you visually inspect a different image. The neural representations of the two different images interfere with each other in your visual maps.

Whereas recollection, working memory, story comprehension, and daydreaming transport us beyond the here and now while we are awake, a different type of imagery transports us while we sleep. Although dreams are not the same as the mental imagery you willfully conjure in the daytime, the two bear a family resemblance. As with imagination and recollection, dreams are triggered by signals from areas of the brain, such as the hippocampus, that use distributed codes to represent events, actions, places, and times. And in all of these cases, those signals act upon sensory and movement brain maps, driving their activity and, in turn, generating your experience of feeling and moving when you are actually doing no such thing. As with mental imagery, dreaming entails activity in brain maps that parallel the contents of the specific dream. For example, neurons within your FFA face zone ramp up their activity when you dream of seeing a face.

The fact that dreams and imagination play out in your brain maps reveals the far-reaching importance of these structures and their zones and boundaries. The very brainscapes that you rely upon to perceive your world also shape and distort the contents of your dreams, memories, and fantasies. This is why, for example, you cannot see out of the back of your head, even when you are asleep. Your visual brain maps devote no territory to representing the space behind your head, leaving no such space for dreams or imagination to occupy.

In fact, the idiosyncrasies of a person's brain maps, such as the relative overall sizes of specific maps, appear to affect the precision of their mental imagery. Because neurons within a large V1 visual map tend to each have smaller receptive fields, people with relatively large V1 maps tend to have higher visual acuity, or the ability to see fine-grained detail. Remarkably, people with larger V1 maps also tend to have greater visual working-memory capacity and better precision for location information in their visual mental imagery.

The foibles of people's brain maps may explain some of the dramatic differences in experienced imagery strength that Galton's respondents reported. But they cannot explain one of Galton's most intriguing observations: some people do not experience mental imagery at all. He wrote, "To my astonishment, I found that the great majority of the men of science to whom I first applied protested that mental imagery was unknown to them, and they looked on me as fanciful and fantastic in supposing that the words 'mental imagery' really expressed what I believed everybody supposed them to mean. They had no more notion of its true nature than a colour-blind man, who has not discerned his defect, has of the nature of colour. They had a mental deficiency of which they were unaware, and naturally enough supposed that those who affirmed they possessed it, were romancing."

Modern studies have upheld Galton's observation: some people simply do not experience mental imagery at all. But why might that be? Science hasn't yet arrived at a comprehensive answer to this question, but we do have intriguing clues. One thing is clear: insofar as the experience of mental imagery depends on activation of sensory brain maps like the V1 visual map, damage that affects a person's perception also affects their mental imagery. When strokes, head trauma, or other misfortunes damage visual areas of a person's brain and disrupt their visual perception, their visual imagery is affected in a similar way. For example, people can lose the ability to see color after brain damage to a particular region of the visual cortex. These patients can still see the world, but it is now in shades of gray. They can also still generate clear mental images, but those images too have been drained of all color. Other people with brain damage that destroyed the FFA face zone lose the ability to recognize the faces around them *and* to generate mental images of specific faces. Even momentarily disrupting neural activity in a healthy person's V1 map with bursts of transcranial magnetic stimulation makes them temporarily perform worse on both tests that require visual perception *and* tests that require mental imagery.

Although damage that wipes out perception tends to wipe out imagery, the reverse is not necessarily true. One example is a patient whom I will call Michael. This sixty-five-year-old man experienced what seemed to be a minor complication during a procedure involving his coronary arteries. Before the surgery, Michael often visualized buildings for his job and visualized faces and events at night before falling asleep. After it, he found that he could no longer visualize images while awake nor see them in his dreams at night. Tests of his visual and neurological function came back normal. He could *see* just fine. He could even remember things that he'd seen. But he could not call the images to mind. As he put it, "I can remember visual details, but I can't see them . . . I can't explain that . . . From time to time I do miss being able to see." Michael's condition—the inability to generate mental images—only recently received a name: aphantasia.

Scientists studied Michael's brain using functional MRI. When he viewed pictures of famous faces, he showed normal activity in the FFA face zone and in visual maps, including V1. But when he was asked to visualize famous faces, these areas showed substantially less activity than they did for normal visualizers who performed the same task. The visual areas of Michael's brain were working and apparently normal, but for some reason they were not being used for imagery. Somehow, the spark that should have set these areas into action had been extinguished.

Whereas Michael lost his ability to visualize imagery after a surgery, others, including Galton's buddies in science, report never having had that ability to begin with. These people are not visually or cognitively impaired. They are perfectly capable of remembering and being creative. Many of them report that they didn't learn until their teens or twenties that other people actually see things in the mind's eye.

All of this suggests that mental imagery is not an essential capacity for human thought. Although many of us harness our brain's machinery for visual perception and co-opt it for remembering, imagining, and mentally manipulating information, the brain can also solve those particular problems in other ways. Mental imagery is simply a neural strategy—one clever trick for helping you wring as many useful capacities as possible out of your finite brain. Your V1 visual map is wonderful at representing visual information in the service of visual perception. Why shouldn't your brain also use it to represent visual information in service of recollection, working memory, story comprehension, and imagination?

The idea that parts of the brain can be co-opted to carry out different but related functions has been called neural reuse. This term makes it sound as if we are taking a discarded, unused thing and finding a new purpose for it, but that is not the case. Rather, we are finding more uses for a thing that is already *quite actively being used.* In that sense, it might be more accurate to call it opportunistic neural cohabitation. In essence, it is like taking a desk that one person is happily using and adding two or three more people to work at the same desk at the same time. If you have a limited number of desks but no shortage of workers, tripling up in this way might be a boon; you are now getting more out of that desk. But of course, there is a downside too, because your three workers are now competing for space and resources. The more productive one worker is, the longer the others are left waiting for tools and space. That is why seeing or hearing something new destroys working memory for patterns or words, respectively; there just aren't enough neural keyboards and staplers to go around.

The fact that mental imagery and perception interfere with each other might help explain why people vary so much in their use of mental imagery. Although there are clearly some benefits to making mental images, there are very likely costs too. By forcing perception to share a workspace with mental imagery, people who make imagery may give up some of their ability to purely and accurately perceive. In that light, aphantasia is not a "defect," as Galton described it. Instead, it is a different way of using the brain, which confers different benefits and drawbacks for perceiving, thinking, and remembering. In short, although opportunistic strategies like mental imagery help us wring new abilities out of our physically limited brains, they are still subject to the stiff competition and tough tradeoffs that characterize every aspect of brain maps and neural representation.

Thanks to intrepid scientists and recent technologies, we now have real purchase on the question "What is mental imagery?" But mental imagery is not the only product of the mind that should, by all rights, lie beyond the reach of science. Another familiar yet ineffable feature of your mind is its capacity to attend. Attention is invisible, elusive, and hard to describe. And yet, like mental imagery, attention operates within your brain maps and can be detected there with brain scans.

HOW ATTENTION WORKS ITS WILY MAGIC

While you have been reading this book, you have been receiving my words and, from them, constructing meaning. There are countless other things you could have been doing instead: noticing the pressure of your shoe against your big toe, hearing the trill of a nearby bird, or pondering that ever-present question, *What's for dinner?* You might now stop reading and do any one of those things, but if you do, it will take something away from my words and their meaning. Some *thing* is bequeathed to either this book or to dinner, or thinly spread like an insufficient condiment across them both. That *thing*, so familiar to us from our youngest years, is attention. Whether out of interest, surprise, or the threat of a failing grade, you invest your attention in one object, activity, or place, only to call in the loan and divert those mental funds to something else entirely. You do this continuously and with ease. You are a veritable Wall Street day trader when it comes to allocating and reallocating this ethereal cognitive resource. But what is attention, and how does your brain make it?

William James, an influential American psychologist of the nineteenth century, described attention like this: "Everyone knows what attention is. It is the taking possession by the mind, in clear and vivid form, of one out of what seem several simultaneously possible objects or trains of thought. Focalization, concentration, of consciousness are of its essence."

Perhaps everyone does know what attention is. But let's imagine for a moment a person or creature who doesn't. Imagine a species from another planet, with brains that record and process all of the sensory details of the world around them faithfully and in their entirety—a sort of panoramic multisensory video recording of every sight, sound, and sensation. To them, James's definition would seem inscrutable. How could our minds take possession of anything? How could they refrain from registering events or objects that are right before our very eyes? What exactly is concentrated? What becomes focalized, and where, and how? The very fabric of their experience would seem to be so wholly different from ours that we might declare it futile to attempt explaining attention.

But let us try. One way to explain attention might be to describe the observable effects that attention has on our behavior. What does paying attention to a place or an object *do* for us? In a nutshell, it makes us better at sensing certain things—things that we are interested in or searching for. We are better able to perceive and detect things—dim shapes, faint sounds, and the like—when we attend to them. We also become faster at finding them. Attending to something is like using binoculars to see distant objects or infrared goggles to see people at night; it boosts the ability to see that *something* is there and to discover *what* that thing might be.

But there is a flip side to attention. In making us better at perceiving a particular target, it leaves us worse at perceiving almost anything else. It seems to be a finite resource that must be strategically moved from one moment to the next like troops on a battlefield. James and many others argue that one reason attention is so useful is because it shields us from the mountains of irrelevant information that our senses would otherwise deliver to us. Without attention, they argue, we would be overwhelmed and overloaded. And yet this wouldn't be true for our space-traveling acquaintance. Our friend is perfectly happy to drink from the fire hose of panoramic sensory experience. There is no reason why a mind *couldn't* perceive and process all of that information simultaneously. And yet it is patently clear that *our* minds cannot. Why is that?

To answer that question, recall the tough tradeoffs inherent in brain evolution. Your brain mustn't be too big or too heavy, or demand too much fuel. Brain maps are one of nature's solutions to this problem – a way to reap the most from a finite brain. But brain maps aren't enough to get you there. And so our brain maps are warped by magnification, bequeathing vast expanses of territory to represent inputs from favored regions of the body or the visual field or the sound-frequency spectrum, while giving short shrift to others. And yet even with warped brain maps, the wide world is *still* too busy and vast for us to manage at any single point in time. If we hoped to receive and process information about all locations and objects and senses from all of our maps all at once, our brains would have to be substantially larger. Implausibly so. Our alien friend could manage this only if he possessed an enormous head, few or imprecise senses, or perhaps all of the above. Attention is a brilliant solution to these constraints. We can have small heads *and* sharp senses, but we can dynamically boost the processing of certain things that are important to us, moment by moment, at the expense of those things that are (for the moment, anyway) not.

It is easy to see attention work its fickle magic on all of the brain maps that we already know and love. When you attend to reading the words written on this page, neural activity in the foveal region of your V1 visual map is high. So too is activity in the zone of your object map devoted to processing letters. But when you shift your attention to how your shoe presses against your right big toe, activity in visual and object maps drops and activity in the big-toe region of your left S1 touch map rises. Do it now; shift your attention from what you see to what you feel. Notice that the same patterns of light and dark were falling on your retinas and the same patterns of compression impinged on the skin of your toes before and after you shifted your attention. The change in intention that came from within you caused a major change in how quickly neurons in your V1 and S1 maps fired, alongside major changes in what you perceived. Just as you can willfully call a mental image to mind by using your brain maps, you can willfully amp up or tamp down the activity in your various brain maps by allocating attention.

Think of how effortlessly and yet profoundly that act changed the nature of your experience – from one that is primarily visual, reflecting a tableau of ricocheting photons, to one that is primarily tactile, reflecting the mechanical pressure of molecules in your skin and molecules in your shoe repelling one another. Now, if you stop to listen to the ambient sounds around you, you will ramp up the firing of neurons in your auditory cortex and simultaneously become aware of sounds – perhaps a clock ticking or a bird chirping – that you may not have noticed before. Tune in to each of your senses in turn. For each, you will find a panoply of sensations awaiting you, unnoticed until you sought them out.

Attention does not just momentarily privilege a particular sense over another. Even within a sensory modality, it can favor the processing of things in certain places. Here too we can see attention at work in our brain maps. For example, when you attend to items at your center of gaze, activity is high in the foveal part of the V1 visual map, regardless of what else might be happening in your visual field. Now let's say that you keep your gaze glued where it is, but you shift your attention outward, away from where you are pointing your eyes. You have probably employed this kind of stealthy attention when you wanted to look at someone without their knowing it. When you shift visual attention in this way, neural activity drops in the foveal part of V1 and ramps up in the corresponding peripheral region of the V1 map.

Attention can also privilege the processing of certain types of things. If you want to locate a face in a complex visual scene, you can attend to faces, wherever they may be. Attending to faces boosts activity in face-preferring zones of the object map, such as the FFA face zone. It will also help you locate the face much faster. Ditto if you want to locate a building in a picture: you attend to buildings and simultaneously boost activity in the PPA place zone. If you want to detect a specific speech sound, you can ramp up corresponding regions of your auditory cortex. If you want to detect whether a fluid tastes sweet, you can increase activity in regions of your taste cortex.

When you zoom in and try to understand exactly how attention affects the firing of individual neurons within these brain maps, the picture becomes substantially more complex. In some cases, directing attention toward a feature or location that a particular cell prefers may simply and straightforwardly increase how often that cell fires. But in many cases, attention has more nuanced effects. It can boost a neuron's sensitivity, making the cell more likely to fire for the faintest glimmer of a target. Or it can turbocharge its firing rate only when that rate is already high—in a sense, amplifying its active state and making the difference between *Nope, nothing* and *Yes, something* more obvious to the rest of the brain.

Willfully attending to something has a remarkable impact on sensory brain maps and actual perception. But what triggers those changes? If attention is like legions of troops being strategically deployed and moved between various senses and parts of sensory maps, who is the general commanding those troops? At this point, we know several key figures in the chain of command—and all of them contain their own maps. Recall the spatial maps of salience or intent in the parietal cortex, which combine information from multiple senses and coordinate systems. These maps collect information from the senses about what and where are important right now. They relay crucial information to the motor cortex, so that you can act upon or react to these important things or people. But they also send this information back to the sensory areas, boosting and hushing regions of the sensory maps as needed in the moment. The motor cortex is also important for directing attention. The frontal eye fields are regions in the motor cortex that contain their own visual map. The neurons within the frontal eye fields generate eye movements and are responsible for directing your gaze toward events and objects of interest in your environment. But they also direct attention-related activity in other parts of the brain, including sensory brain maps.

Although there is no single area of the brain that plays puppet master when it comes to directing attention, the major players are poised at the intersection of action and perception. Attention knows what is relevant for behavior at this moment in time because it is listening to your motor system. That is the beauty of your interconnected brain: perception is always shaping action even as action is perpetually, fundamentally shaping perception.

Of all the thoughts and capacities that arise from the mind, attention and mental imagery seem the most personal and ethereal. As such, it is astonishing how much we now understand about what these phenomena actually are and how they play out in our brain maps.

Another grand challenge for neuroscientists has been to understand how the brain represents concepts and meaning. Abstract concepts like number, time, love, and failure are divorced from the here and now. They cannot be seen or touched, but they still must be represented, understood, and discussed. Remarkably, brain maps have a role to play when it comes to representing these intangible notions as well.

10

Comprehending and Communicating with Brain Maps

BY THIS POINT IN OUR JOURNEY through brain maps, one thing is patently clear: your brain maps are grounded in your physical body and your physical senses. They inform you about the crucial physical properties of the world around you. The common thread among them is how very physical they are, each in its own way. Even when we consider the most mental of phenomena—mental imagery and attention—we find that they mimic physical sensations and actions within brain maps and can be physically detected there.

Yet so much of what we think and talk about is abstract and perceptually impenetrable. Just try to touch debt or compound interest. See if you can hold time in your hand or put love in your pocket. Many of the concepts that rule our financial, social, and emotional well-being have no shape, no color, no odor or weight. How do we grasp these intangible concepts? As it turns out, we often do so by aligning abstract concepts with physical dimensions and then putting our brain maps to work.

FASHIONING NUMBERS AND TIME FROM FINGERS AND SPACE

On its face, mathematics seems like the pinnacle of human abstraction. What in real life might be three specks of dirt or three galaxies of shimmering stardust can both be represented by one ambiguous symbol. Whether we are talking about a household budget, an asteroid's path, or the spread of a virulent new illness, we can represent, compute, and predict using the same simple lexicon of numbers and mathematical operations. Although mathematics may be harnessed to describe things in the here and now, it is very much beyond both here and now. The number 3 is a symbol, but you will never touch or see the abstract concept of 3 in the physical world. How do brains represent such a thing?

The root foundations for our understanding of numbers arise from at least two brain-based abilities. One is an approximate number awareness. This ability to approximate *how many* is not unique to humans; most species need to be able to roughly size up how many things they encounter. How else could a creature run toward the tree that bears more edible fruit or away from a space occupied by more predators?

Yet this approximate number sense is inexact and quite limited. It permits us to discriminate between only small numbers (1 versus 2 or 3) with any precision. As the numbers go up, precision goes down. As far as your approximate number sense is concerned, 6 and 7 are more or less the same, and larger numbers like 12, 14, and 20 are identical, with an approximate value of *a lot.* This number sense will never get anywhere close to representing larger precise numbers like 1,109. Nor can it perform even basic computations, such as 14 plus 18 equals 32.

Neuroscientists have identified brain maps that support this approximate number sense in monkeys and in humans as young as three months old. In adult humans, functional MRI brain scans have revealed six maps of approximate number in each hemisphere of the adult human brain—most of them in the parietal cortex. These are smooth, continuous maps of quantity with approximate number as their organizing dimension. Each map ranges from representations of one to approximately four, five, six, or seven (that is, a bunch).

On top of this approximate notion of quantity, we construct associations between numbers and space. These associations are not limited to mathematicians or even humans capable of doing basic arithmetic. By means of a variety of tests, we know that birds, monkeys, and human infants demonstrate that they make associations between quantity and distance or space. Yet for humans, these associations become stronger and more consistent over the course of schooling. The associations also tend to be culturally specific. For example, people from cultures that read from left to right strongly associate small numbers with space on the left side of the body and larger numbers with space on the right side. That's because their exposure to written numbers in school and elsewhere involves arrangements of numbers that increase from left to right. A quick look at the time lines and the horizontal axis of nearly every graph presented in school curricula reveals that this convention for arranging numbers is pervasive. But children who grow up in cultures that read text from right to left associate space and number in the opposite way; they tend to associate smaller numbers with space on the right.

That children learn these conventions is not surprising. What is surprising is how their brains adopt these spatial associations and integrate them into representation of number. This involves co-opting brain maps that represent physical space and using them to represent number. The resulting overlap between brain representations of space and number has been called the mental number line.

Yet we do not align numbers and space in one single way. Any person can employ more than one space-number association. For example, people demonstrate that they use a vertical mental number line to associate small numbers with lower regions of space and large numbers with higher regions of space. Whether we think about numbers using a left-right or low-high association depends on the task and the context.

The best way to see how intertwined space and number representation are in the mental number line is to witness how they interfere with each other. For example, seeing small numbers at the center of their visual field biases people toward directing their spatial attention to the side of their visual field that they associate with smaller numbers. People are also faster at judging whether a large number is even or odd if it is displayed on the side of their visual field that they associate with large numbers, rather than the side they associate with smaller numbers. And what is considered a large number or a small number also depends on context. If you are considering how many scoops of ice cream to order, one scoop might be associated with the left, two scoops straight ahead, and three scoops to the right. But if you are comparing car prices, even the numbers represented to your left and thus considered smaller numbers might be in the tens of thousands. In short, we can use our mental number line to flexibly compare and reason about exact numbers, even very large ones that are far beyond the reach of our approximate number maps. And because we co-opt spatial maps to do this task, thinking about numbers affects how we allocate spatial attention, just as attending to regions of space affects how we think about numbers.

A dramatic demonstration of this association comes from patients suffering from a condition called hemispatial neglect. Hemispatial neglect typically occurs after a stroke or other traumatic event causes damage to the parietal cortex on the right side of the brain. Recall that the parietal cortex is important for guiding attention toward the space and objects around you. Just as the visual cortex on the right side of your brain represents the left half of visual space, parietal maps in your right hemisphere represent targets of attention in the space around the left side of your body. Patients who develop neglect after damage to the right parietal cortex show a profound and ongoing inattention, or lack of awareness, for objects, events, and even their own body parts on their left side. When presented with a platter of food at mealtime, someone with hemispatial neglect might eat all the food on the right side of his plate and then ask for a second helping, unaware that the plate is still half full. But if someone kindly rotates the plate 180 degrees, so that the remaining food is now on the patient's right side, he will now be aware of it and can happily polish off the rest of the meal.

Remarkably, hemispatial neglect also affects how a patient performs on a simple number task. Let's assume we are observing a patient who is from a culture that reads from left to right. She is seated in front of a computer and is shown single digits, one at a time, at the center of the screen. For each digit she is shown (1 through 9, excluding 5), she is tasked with indicating, as quickly as possible, whether that digit is less than or more than 5. Although the patient will be able to give the correct answer for all tested digits, she will be slower to give a response when the digit in question is 4 than when it is 6. Why is that? Because to do the task, people imagine themselves at the 5 position of a mental number line. This turns the numerical problem into a spatial one: is each presented digit to my left or to my right? Yet because of our patient's hemispatial neglect, she cannot attend to numbers to her left and therefore can't discern that 4 is to the left of 5. To answer the problem correctly, she has to use another strategy, which takes extra time.

In addition to the strong evidence that our brains use space to think about numbers, they may also use our own awareness of our bodies for that purpose as well. The first hints of this idea came in 1924 when Josef Gerstmann, a distinguished Austrian neurologist, described a patient who experienced unusual symptoms in the months after she suffered a stroke. Gerstmann observed that the woman showed a "marked inability to handle numbers and comprehend them. Serious impairment of the ability to calculate." She also showed "an isolated impairment of her ability to recognize and select her individual fingers. She is unable to differentiate her forefingers, middle and ring fingers, small fingers or thumbs in any given moment, and looks disoriented about them. When instructed to touch or indicate a specific finger of hers, stretch it, or name it, she makes repeated errors, and shows the typical helplessness that is associated with unawareness or embarrassment." When similar impairments were observed in other patients, scientists discovered that damage to a specific region of the parietal cortex was to blame.

Gerstmann ventured a guess as to why capacities for finger awareness and calculation might be linked within the brain. He noted that children often learn to count and do basic arithmetic with their fingers. The link between fingers and numbers is so tight that most cultures use base-ten number systems that group numbers in units of ten, and many number words, including *digit,* come from terms originally used for fingers or hands. If children and cultures use fingers to enter the realm of quantification, maybe the brain does as well.

Decades later, the scientists Elizabeth Warrington and Marcel Kinsbourne, then at the National Hospital, Queen Square, London, studied patients with parietal damage and symptoms like those of Gerstmann's patient. They described in rich detail the patients' struggles with comprehending their own fingers. They wrote, "The fingers have lost their individuality not only as regards their serial order, but in a more intimate sense. Their very differentiation from one another, at least in terms of touch, the tactile finger schema, is lost. The fingers are as if fused into a solid lump." This lost organization may in fact be the loss of a brain map that aligns tactile and spatial coordinates on the body and orients body-centered attention with respect to the fingers and hand.

Recent studies using modern technologies have supported the idea that a region of the parietal cortex contributes to finger awareness and number processing. One study used transcranial magnetic stimulation on healthy participants to temporarily hush the activity of neurons in this region. After stimulation, the subjects performed poorly on tests of finger awareness and number processing. Another research group stimulated this parietal region directly with electrodes while patients were undergoing surgery to remove brain tumors. When the scientists stimulated this region of the brain, the patients had problems with finger awareness and calculation. One patient who was given a simple arithmetic problem during this stimulation asked the physicians to repeat the numbers in the problem many times because he said that he could not understand it.

Experiments using functional MRI brain scans have also found activity in and around this region of the parietal cortex when healthy participants distinguish between their individual fingers and perform numerical calculations. But scientists note that performance of these two activities may coincide with brain activity in two distinct yet neighboring bits of the brain, rather than identical terrain. Instead of using a multisensory spatial finger map to make calculations, we may reapportion some of its territory to use for calculation, much in the same way that we carve out some of the FFA face zone in the left hemisphere and begin using it to recognize letters when we learn to read.

The link between fingers and numbers is also bolstered by studies of mathematical learning. Children's performance on finger-awareness tasks is a particularly good predictor of their performance on numerical tasks up to three years later. It may be that children who are better able to mentally represent and deploy attention to their individual fingers are better equipped to use finger-based attention or mental imagery to represent numbers or calculations. This would help them make the leap from doing math on their fingers to doing math in their head.

At the same time, it is clear that counting on fingers is not a prerequisite for learning to do arithmetic. For example, blind children do not tend to count with their fingers, yet they have no impairments in processing numbers and carrying out calculations. Likewise, children born with cerebral palsy, who have trouble moving their hands, tend to have poor finger awareness, yet they perform fine on arithmetic problems. In short, it seems as though counting on one's fingers is a useful and efficient strategy for learning to represent and manipulate numbers, but it is by no means the only avenue to numerical competence.

Although there remain many outstanding mysteries about our capacity to calculate, two themes are already apparent. First and foremost, it is clear that we do not represent an abstract concept like the number 3 in an abstract way. Our ability to comprehend and compute relies upon a profoundly physical foundation, capitalizing upon or appropriating brain maps for physical space on and around our bodies in order to think about numbers.

A second theme that emerges from studies in this area is that there are many different ways to think about number. The culture within which children learn about numbers affects how they associate numbers with physical space and, consequently, how their brains harness maps to support computation. Your representations of number, space, and parts of the body are interwoven in ways that reflect your culture, your language, and the situations you find yourself in. A school-age child in America learns to count on her fingers from left thumb to right thumb, whereas a child of the Yupno people of Papua New Guinea learns to count on her fingers from left pinky to right thumb, then on her left toes, then right toes, then her eyes and ears, followed by other landmarks on her face and body. Children from both cultural backgrounds can compute a simple sum and generate an identical answer, but precisely how they think about the numbers and reach the solution may be different indeed.

Time is another abstract concept—something you cannot see or touch but that you probably depend upon even more than you do the concept of number. You need to know when things happened to understand why they happened or to make predictions about what might happen next. You need to know how long certain things have been happening so that you can estimate when they will end, and prepare accordingly. And so you and I and everyone else have to devise ways of representing and thinking about time despite its invisible and intangible nature.

Just as our concept of numbers originates with an approximate number sense rooted in the parietal cortex, our concept of elapsed time, or duration, may begin with parietal representations of approximate length, size, or volume. That's because we tend to use spatial metrics like size or distance to estimate the passing of time. For example, you can estimate how much time it took you to walk somewhere based on how much distance you covered in that time. Or you might estimate how long you have been pouring water into a jug based on the distance between the top of the water and the bottom of the jug.

As we do with number, we acquire culture- and context-specific associations between time and space. These associations allow us to keep track of and communicate information about time. Just as we use a mental number line, psychologists have shown that we use a mental time line as well. Often these mental time lines are horizontal. In cultures that read from left to right, the past is thought to be on the left and the future on the right. In cultures that read from right to left, the order is reversed, and the future is thought to be on the left. But these time lines are flexible and depend on context. For example, if the task is to determine whether a date occurred in the past or will occur in the future, you might assign the center of the time line to the present day. But if your goal is to determine whether dates happened before or after a noteworthy past event, such as the start of a war, you might center your time line on that event, so that you represent the more distant past to your left and the less distant past to your right.

Horizontal time lines are not the only kind of mental time lines people use. In English-speaking cultures, another prominent mental time line is back-to-front. *Now* is represented as here (where you are standing), with your future stretching out before you and the past trailing behind you. The Aymara people of the Andes have the opposite association: for them, the past is before them and the future is at their back. Native speakers of Mandarin have a vertical mental time line that associates earlier events with upper space and later events with lower space. The Yupno people describe time according to their local landscape: the past is downhill, toward the mouth of the local river, whereas the future leads uphill, toward the river's source. As is the case with mental number lines, context matters for mental time lines. For example, people who fluently speak both Mandarin and English are more likely to arrange events according to a vertical time line when the instructions are given in Mandarin rather than English.

Just as hemispatial neglect revealed the neural foundations of the mental number line, it can offer insights into the basis of the mental time line. Specifically, people with hemispatial neglect have problems with using their mental time line to reason about temporal order. In one study, patients heard stories about fictional events that either happened ten years in the past or would happen ten years in the future. Later, they were reminded of specific events and asked to say whether they happened in the past or in the future. In cultures that read from left to right, people do this task using a mental time line that aligns the present time with their location in space, so that past events are represented to the left and future events to the right. As we know, patients with hemispatial neglect do not attend to the left side of space. Although they correctly identified the future events, they incorrectly identified most of the past events as future events as well.

Like many examples from the annals of neurology, the profound cognitive difficulties experienced by patients with hemispatial neglect may seem exotic or even bizarre. But through their example, they lay bare something that is truly bizarre within each of us. To think about things beyond the here and now, we hijack the very brain maps that support life in the here and now. We recruit and reuse them, harnessing one ability in the service of many others, so that we can expand our mental horizons without expanding the size of our skulls. We rise above our senses and our bodies, but only by opportunistically building upon the cognitive foundations they have laid down for us.

DECIPHERING MEANING

Number and time are essential concepts for any growing child to learn. But they are just two concepts among many that guide our lives and determine our well-being. Risk, conflict, love, and jealousy are just a few of the many others that cannot literally be seen or touched but that nonetheless can be felt through their influence on our relationships and fortunes. As intangible as such concepts are, we have no problem understanding and talking about them. In fact, they tend to be favorite topics of discussion at the water cooler or dinner table.

How do our brains represent concepts and forge an understanding of them? And once we achieve this understanding, how on earth do we manage to share it with others?

Up to this point in our journey, we have looked for answers to our questions within individual brain maps. We have considered their organizing dimensions, their distortions, and their roles in supporting and shaping our ability to perceive and to act. But brain maps cannot function in isolation. If excised from your brain, your primary visual cortex would no longer contain a map. It wouldn't even be *visual.* It would just be a sad chunk of dying flesh. What makes V1 a visual map is the kind of information that flows into and out of it and the pattern of connections through which that information travels.

The interdependence of maps becomes all the more apparent when we consider the neural mechanisms of thought beyond sensation and action. Think about the many maps called into service when you estimate or reason about time. There are sensory maps, like V1 or A1, that initially process the sensory events that you will need to consider in relation to time, like seeing and hearing what happens when you put a pie in the oven. There are visual and parietal maps that process number and space and parietal and frontal maps that guide spatial attention within a mental time-line framework. To estimate how much time has elapsed during events or activities, we also engage a movement area in the prefrontal cortex and two structures that lie far beneath the surface of the brain. All three of these regions are involved in generating action, and all three contain their own varieties of maps. At a glance, then, we are talking about sensory maps, multisensory salience maps, and movement maps throughout the brain, all of them working together to support this singular notion of time. But even that description is not sufficient because regions like the hippocampus, which contain distributed codes rather than maps, play their own essential role in supporting our representation of time.

Throughout this book, we have seen how damage to specific brain areas can wipe out key abilities, like John's ability to recognize his wife's face or Gerstmann's patient's ability to do basic calculations. It is tempting to imagine that these abilities reside in their respective maps. But that is not how brains work. In the case of V1, for example, other regions of your body and brain are feeding it information—maybe details from your retinas about incoming light or details from the front of your brain about what you are expecting or hoping to see. Other regions, such as your object map and the face, place, and body zones within it, are receiving information from V1 and using it to help you recognize and eventually act upon what you see. In this way, your brain is like a complex subway system with many routes and transfer points. If the track has been damaged at a particular station, the lines that run through it will be shut down. But although the line needs that station to complete its route, the line does not reside in the station. It does not "reside" anywhere because it depends on motion. It can serve its purpose only if the stations are all open, so that the trains keep passing through.

To understand understanding, think of maps as crucial stations in a larger system. Meaning does not reside in any single map, but many maps can work together to create meaning. When you talk with friends, drive a car, or watch a movie, your brain is churning with activity. Meaning is culled from the collective action of many cells in many maps at once.

Neuroscientists have glimpsed intriguing echoes of this constructed meaning in individual brain maps. For example, reading the word *kick* weakly ramps up activity in the leg region of the M1 movement map, whereas reading *lick* weakly activates neurons in the M1 face region. The neural activity in your movement maps when you read words related to a specific physical action resembles a weaker version of the activity in those maps when you actually do engage in that action. Likewise, when you watch someone else act in a certain way, several of your frontal and parietal maps will display patterns of activation as if you were taking that action yourself. The same principle applies to brain maps representing information that comes to the brain from our senses. Watching someone else pet a dog or hearing someone else tap on a drum activates your touch maps. Reading a word that is conceptually related to sound, like *telephone,* activates regions of your auditory cortex; reading taste words like *salt* activates your taste cortex; and reading smell words like *cinnamon* activates your olfactory cortex.

Hundreds of studies using functional MRI brain scans have revealed that our movement and perceptual maps can tell us more than simply what we are doing or sensing at the present moment. When we observe a word, an object, or an action, the motor, sensory, and spatial regions of our brains begin reflecting information about the meaning of what we have seen, heard, or encountered – all in a matter of milliseconds. This is true when we think about concrete things like cinnamon, and likewise true when we think about abstract things like number and time.

Given the challenges of peering into the living, breathing human brain, we can't tell for sure whether the same exact neurons are involved in, say, helping you move your leg *and* process the meaning of the written word *kick*. It might be two separate sets of neurons forced to share a studio apartment. Either way, there is a link. It is tempting to say that this link is the stuff of meaning – that mentally simulating the act of kicking is how you know what kicking is. But this explanation can't be right. A person who has been quadriplegic from birth and therefore has never kicked anything can still know perfectly well what kicking is and might even spend her spare time cheering soccer teams. Doing can be related to knowing, but they are not the selfsame thing. Likewise, I can know that astronauts float in space, although I have never floated free of gravity nor been in space myself. I know that elephants are heavy, although I have never lifted one. And I am quite certain that one day I will die, although I have never (to my knowledge) been dead before.

Still, my collective experiences with an object, person, or action shape and enrich how my brain will represent it. If I have seen people kick but have never kicked anything myself, I might represent *kick* more with visual maps and less with movement maps; if I have played on soccer teams for years, that representation might be quite different. In either case, I will understand what *kick* means, but my different experiences and their sensorimotor details will become integrated into what that action means to me—and how my brain represents it.

Scientists are still figuring out how these details of experience become integrated and stored in the brain, so that the next time I think of kicking or read the word *kick* on a page, these facets culled from experience can return to my mind and occupy my brain maps. A brain area near the tip of the temporal lobe may play a special role in supporting multisensory representations of meaning, tying together, say, how kicking feels, how it looks, how one does it, and what it is called. When a degenerative disease causes neurons in this region to die, people develop a condition known as semantic dementia. This condition is characterized by an inability to think of particular words, an inability to understand words used by others, and a loss of general knowledge about people, objects, actions, feelings, and other concepts. Unlike John, who lost the ability to visually recognize a carrot but retained all of his knowledge about how carrots are grown and consumed, a patient with semantic dementia might shrug and say, "I'm not sure what a carrot is. Maybe something you eat?"

Taken together, it is clear that representing any particular meaning entails marshaling many areas of the brain and its maps all at once. How, then, do you share this meaning with someone else, so that two brains represent the same meaning at the same time? Shared meaning is essential to communication. But what *is* it, and can we observe it in the brain?

We've seen how the nature of your prior experiences with an object, action, feeling, or concept precisely affect how your brain represents it and therefore how your sensory, motor, and spatial representations of that object, action, or concept will be different from mine. Still, our experiences should be similar in many ways, owing to the fact that we both live on Planet Earth and have similar body layouts, senses, and physiological needs. My heart pounds when I run, just as yours does. My lungs expand to draw breath, just like yours. If we come from different cultures, there will undoubtedly be differences. For instance, for different people, eating might involve scooping with bread, plucking with fingers, or maneuvering with utensils such as chopsticks. Those experiences are truly varied. But the sensation of chewing and swallowing, and the relief of quelling a fierce hunger—those will be the same. This raises an intriguing question. You and I have similar brain maps. We have also had many similar experiences with objects, actions, feelings, and concepts. Does that mean that your brain and mine will represent concepts in a similar way?

In recent years, ambitious studies have used functional MRI brain scans to look for shared signatures of meaning across human brains. One way to do this is by engaging people with a similar meaningful experience: watching the same movie. A pioneering study used functional MRI brain scans to see what happened throughout people's brains while they viewed the classic western *The Good, the Bad, and the Ugly.* At any point in the movie, the patterns of activity observed in a particular person's brain were remarkably similar to the patterns observed in the brains of other people watching that point in the movie. Activity in nearly a third of the surface of each viewer's brain was synchronized with activity in other viewers' brains while they watched. This synchronized activity occurred primarily in the back half of the brain, encompassing maps in the occipital, temporal, and parietal cortex.

Another experiment used a similar movie-watching design to take these ideas a step further. The scientists analyzed how patterns of activity common across different people's brains corresponded to the specific types of objects, events, and actions being shown on the screen. Once they characterized these common patterns, they used them to decipher activity in individual brains. Specifically, they attempted to determine which kinds of things—say, a car, a tree, or an athlete—each test subject had been watching on the screen at any given moment during the scan. Using the remarkable similarities between people's brain activity, the scientists were able to correctly guess what individual people were watching 76 percent of the time.

In a sense, these studies were describing how the brain, including its resplendent maps, divvies up responsibility for processing and representing different concepts as they come and go in the context of the movie. Sensory, multisensory, and movement maps were all involved. For instance, watching two people shake hands in the movie might, in that moment, activate hand and grasping regions of movement maps in the frontal and parietal cortex, hand regions of the touch cortex, face and body zones within the visual object maps, and so on. That doesn't mean that all of these areas are uniquely dedicated to processing the action of people shaking hands – in fact, none of them are. But all of them are marshaled in service of processing and representing this action. And they are marshaled not just in your brain, but in my brain, and in other people's brains too.

Why is the global layout of activity similar in your brain and mine? Because our brains developed according to the same basic genetic and chemical blueprints, creating the same basic brain architecture nestled within the same basic body architecture. Because we share the same senses and feel the same tug of gravity and inhabit bodies that work the same way. Because when you opened your newborn eyes, you saw a face staring back at you, just as I did. Because you heard a language and learned to speak it with your lips, tongue, and teeth, just as I did. Because our sun rises and falls with the same rhythm over distant horizons. In many ways, we live the same life, you and I. And in many ways, we share the same brain. And that simple fact, that likeness between us, makes all the difference in the world when it comes to sharing meaning across brains.

Pioneering studies using functional MRI brain scans have lifted the veil on mutual understanding. They have shown that when people communicate, their brains synchronize. When you listen to someone spontaneously tell a personal story from the past, the activity in your brain maps resembles, moment by moment, the brain activity of the person telling the story. For most of the areas in your brain, this activity will lag behind the speaker's by one to three seconds, reflecting the time that it takes you to hear and process the speaker's words. The more your brain activity resembles that of the speaker, the better you will comprehend the story and will be able to report it correctly later. This synchronized activity between the speaker's and the listener's brains specifically reflects mutual understanding. If the speaker tells the story in a language that the listener does not know, this synchrony does not occur.

Synchrony is not just important for understanding other people's stories. It is also important for education and learning. When a teacher delivers a lecture to a class of students, activity in vast swaths of the students' brains synchronize with that of the teacher's brain. And the stronger this synchrony is, particularly in key parietal and visual maps, the better the student will perform later on tests of the material. In short, our shared experiences, enshrined in shared brain patterns, help make shared meaning and shared knowledge possible.

This link between brain synchrony and shared meaning suggests a new way to think about communication. Consider the following real-life example of communication across time and space. From his boarding school in England in 1924, the eleven-year-old Alan Turing sat down with a fountain pen of his own invention to write a letter to his parents, who lived in India. It was just one of many letters he wrote to them, recounting the goings-on of his young life:

> Dear Mother and Daddy. I have started writing with my fountain pen again, please tell me if you think my writing is worse with it . . . I do not know whether I told you last week but once when I said how much I hated tapioca pudding and you said that all Turings hate tapioca pudding and mint-sauce and something else I had never tried mint-sauce but a few days ago we had it and I found out very much that your statement was true.

In writing his letter, young Turing shared an event: how, through a single unpalatable experience, his impoverished knowledge about mint-sauce (based on hearing about it and perhaps seeing it) turned into a richer knowledge of what mint-sauce is and how little he cares for it. To share this event, he used his handmade pen to convert his thoughts into ink marks on a sheet of paper—a format durable and portable enough to travel across continents to his parents' waiting hands. He would have had no reason to question that his parents would unfold the letter and promptly decipher his ink marks, converting them back into meaning—*his* meaning about *his* recent run-in with mint-sauce. From start to finish, the process would have seemed mundane. There is nothing technologically challenging about writing or reading a letter.

The same could not be said fifteen years later, during World War II, when Turing was recruited for a secret mission to help Britain and its allies crack the Germans' secret wartime codes. The Germans used a system called Enigma to encrypt and decipher messages. The system was considered unbreakable, allowing the German military to broadcast their encrypted messages by radio. They knew that the Allies would also receive the signals, but that was no cause for alarm. An encrypted signal by itself is meaningless garbage. It becomes a message only if the person receiving it has the key to decoding it.

The Enigma system involved substituting an encryption letter for each letter in the actual message. The letter substitution pattern was complex and ever-changing, so that the message letter A might be encrypted to the letter X in one instance and then to the letter E later in the same message. The trick to encrypting and decoding the messages lay in a physical device called the Enigma machine, which looked like the offspring of a typewriter and an old telephone switchboard. Any properly set Enigma device was equally capable of translating any message into encrypted code *and* translating any encrypted code back to the original message.

To encrypt a message, an Enigma operator would type each letter of the message, one key at a time. For each key press, a small light bulb would illuminate its substitute—the encryption letter. For example, someone hoping to encrypt the message *RETREAT* would have typed R-E-T-R-E-A-T on the keyboard and wrote down the letters that were illuminated in turn—say, Y-P-O-B-R-Z-I. The encrypted message, *YPOBRZI,* could then be broadcast over the radio. An Enigma operator on a U-boat receiving the message would write down *YPOBRZI* and then sit down to his device. Assuming the settings on the receiver's Enigma device were the same as those on the sender's, the receiver simply had to type Y-P-O-B-R-Z-I into the machine. For each encrypted letter he entered, a small light bulb illuminated the decoded letter. In this way, letter by letter, the original message, *RETREAT,* would be revealed.

Polish mathematicians and French spies kicked off the laborious process of cracking Enigma's code. Turing and his fellow mathematicians in Britain built off their work and eventually, through a combination of innovation and perseverance, were able to decipher the German messages and turn fortune in favor of the Allies.

Although Turing's work in cracking the Enigma encryption system had higher stakes than his letter writing as a boy, both were ultimately a matter of codes. Codes convey information using arbitrary symbols that can be transmitted in writing, through sound, and even through touch, as it is with braille. Codes can successfully transmit a message, provided that the sender and receiver both know the key to converting the message into symbols and back again.

In the case of language, it is easy to think of the brain as a souped-up Enigma machine. It takes meaning and converts it into spoken or written language. It can also do the reverse, converting language back into the original meaning. From this point of view, the meaning is the message, the language is the code, and the brain is the hardware that translates between the two. But given the importance of shared neural activity in generating shared meaning, we might consider an alternative: activity within the brain, or the brain's state, is the message itself. And the key to unlocking the code, to enabling communication between one human being and another, is the common brain architecture and common physical experiences that help ensure that our similar brains are imbued with similar meanings.

When the young Alan Turing sent his sweet message to his parents five thousand miles away, he wasn't just sending them words. He was sending them a series of brain states—patterns of activity across visual, auditory, taste, smell, touch, space, grasp, and other maps that represented pens, ink, handwriting, disgust, tapioca, the Turing family, truth, and yes, mint-sauce. Thoughts are contagious. And they are contagious for the very same reason that illnesses are contagious—

because you and I share the same body. A microbe and a notion can make themselves equally at home in you and in me because we are, despite our differences, virtually the same.

This way of thinking about meaning has implications for artificial intelligence and the prospects of communicating in a meaningful and unscripted way with computer programs. Over the course of his short yet dazzling life, Alan Turing famously considered whether machines can be intelligent and specifically think and communicate like humans. This question resonates today. If the human brain is merely a translation device that converts words to meaning and back again, certainly a computer program can accomplish the same. Yet we run into a stickier problem if meaning is generated and represented through our experiences in a physical world with a physical body. A disembodied program suffers none of the constraints that we do. Its processing capacity is not limited by the need to keep heads small and appetites manageable. But it also cannot feel hunger or taste tapioca or grasp a pen, or even experience anything remotely like anything the slightest bit like those things. It would seem to be unfettered in all the wrong ways.

Let me be clear: I am not saying that machines can't be intelligent. Computer programs already best us on matchups of information processing and storage, pattern recognition, and games of strategy. When we talk about whether there is intelligent life on other planets and whether we will see true artificial intelligence on earth, we overlook the nonhuman biological intelligence that surrounds us and the artificial intelligence that is already recognizing our faces, organizing our schedules, and storing our collective knowledge. What, then, are we waiting for?

What we mean by intelligence—what we have always meant by intelligence—is thinking the way humans do. Generating and sharing meaning like we do. It is succeeding at the particular accomplishments that we value and viewing the world from the same vantage point that we view it from. An intelligent entity is something that we can communicate with to achieve mutual understanding. But the problem with that definition is apparent. We understand the world in terms that are deeply rooted in our physical experiences – in the abilities and limitations of our particular bodies and senses, the idiosyncratic properties of this particular planet we inhabit, and above all, our needs and opportunities for staying alive.

The question is not whether we can create artificial intelligence, but whether we can create an intelligence that understands the world as we do despite the absence of our physical bodies, our needs, and our environments. To us, a concept as simple as "house" goes beyond an edifice that provides shelter. It is a context, an environment, a landmark, a starting point, or a destination. We know a house by its appearance, but also by its textures, its sounds, and maybe even its smells. Some of your notion of what a house is will be different from mine, but some of it will be the same. You and I have both felt houses with our hands and traversed them with our feet. We have sought shelter in them. We have walked around them and inside them countless times, but rarely if ever would we walk under or over them. We do not swim or fly through them. We do not pick them up, and they do not pick us up in return. No one has ever had to teach you these features of a house. But if there comes a day when you can have meaningful and not absurd conversations with a computer, it will be because someone taught it all of those things, and more. And if it turns out that we never manage to have real heart-to-hearts with a computer, it will not be because we possess an unmatchable intellect. It will be because of our deeply physical existence and the boundaries that it places upon our own understanding of the world.

Achieving mutual understanding is no small feat. It is probably one of the most amazing things we do. Our alikeness – physically, perceptually, and neurally – makes that process possible, or at least a lot easier. It allows me to reinstate in my brain much of what was happening in your brain simply by listening to you speak or watching you move and interact with objects and people. In a sense, this reinstatement is a kind of mind reading that we automatically engage in all the time.

Yet science's growing knowledge about brain maps opens up opportunities for new kinds of mind reading. Just as Jackson's discovery of the M1 movement map allowed MacEwen to find and excise tumors more than a century ago, scientists are now harnessing scientific knowledge about brain maps to make technological leaps once thought impossible. From restoring movement in people with paraplegia to communicating with people trapped in a vegetative state, these advances may empower people sidelined by brain damage or illness, even as they spark difficult questions about personal privacy and sovereignty.

11

Maps as Portals: Mind Reading and Mind Writing with Brain Maps

Knowledge about a representation is a powerful thing. That's because once you know *how* something is represented, you can eavesdrop on or manipulate *what* is being represented. That is the power now open to us when it comes to the human brain. Scientists, physicians, and corporations have the knowledge and technology required to perform at least some forms of mind reading. About that fact, there is no dispute. But what does it mean from a practical standpoint, and should we be concerned?

The answers to these questions hinge on three factors. The first is the forms of mind reading that are actually possible, now or in the near future. In the coming pages, we will tour several ways in which brain maps already serve as portals, enabling reading from or writing to the brains of others. These technologies are broadly called brain-computer interfaces, or BCIs. We will see how the balance of both shared and unique features in your brain maps makes some forms of mind reading and writing possible already, whereas others will probably remain permanently out of reach.

A second factor is practicality. Even if a technology is possible, it may not be practical to introduce it to a mass market. For example, Ford unveiled its Mach 1 Levacar more than sixty years ago. For decades, it seemed that we were on the cusp of commuting to work in flying cars. Surely, if we could fly to the moon, we could fly to the grocery store. And yet our personal Levacar never materialized. Technologies can emerge and thrive only if the physical challenges are surmountable, the required materials affordable, the consumer interest substantial, and the alternatives scarce. It's a good thing we didn't spend our time devising a three-dimensional traffic system for our Levacars and Sky Commuters; hand-wringing about technologies that may never see the light of day could keep us very busy indeed. And yet if a technology *is* practical and truly on the verge of transforming our society, we would be foolish to ignore it. Only by anticipating emerging technologies can we have a say in how our world will be altered by them. As such, it is important to examine whether mind reading and writing technologies are practical and ready for the general public.

A third consideration is the potential consequences of the technology. In this case, what purpose are the technologies under development intended to serve? And what sorts of secondary repercussions might they have on our health, privacy, and autonomy? I will turn to that topic next, considering what we stand to gain and what we may lose if these technologies become woven into our everyday lives.

From the outset, it is important to acknowledge the opinions, fears, and fascinations that many of us already have regarding these concepts. *Mind reading* is a loaded term. A mainstay of sci-fi dystopias and paranoid delusions, it often suggests victimization and an intrusion into or theft of a precious and private possession: one's thoughts. But in fact, most work on developing mind-reading technologies is meant to empower people who are ill or who have been sidelined by a damaged brain or body. This kind of mind reading focuses first and foremost on the wishes of the person it is intended to help. In other words, the mind that is being read *wants* to be read. This simple but important fact stands in contrast to our deepest fears about what it means to read the minds of others.

To appreciate the unique opportunities mind reading can hold to help the powerless, we need look no further than a twenty-three-year-old woman I will call Carol. Carol had not spoken since the day she tried to cross a busy street and was hit by two cars. The force of the collision profoundly damaged her brain, stretching and bruising precious tissues in her frontal lobes and elsewhere. Her brain swelled dangerously in the hours and days after the accident. She survived, but afterward she lived a still and silent life in what doctors call a vegetative state, unresponsive to commands, to sounds, to the people around her. She lay in her hospital bed, month after month, without showing even a glimmer of awareness about her surroundings.

Like others who have fallen into a vegetative state, Carol did not give observers any reason to think she was conscious, much less able to think or follow instructions. But Adrian Owen, a neuroscientist then at the University of Cambridge, wondered if he could use functional MRI brain scans to find out whether Carol and others like her were in fact conscious. This goal presented an unusual challenge: there is no single line that divides people who are conscious from those who are not. You might use functional MRI brain scans to test whether a patient's V1 visual map is activated when she is shown colorful pictures, or if her A1 sound-frequency map is activated when music is piped into her headphones. But finding such activity in the sensory cortex offers no conclusive answer as to whether the patient is truly aware of her surroundings. For example, Edgar Adrian could eavesdrop on the neural messages in the Shetland pony's S1 touch map even while the animal was anesthetized, its limp body propped up by a bench and some sandbags. In the context of a fully functioning brain, activity in the primary sensory maps can be pivotal in shaping one's conscious awareness of sensory events. But in the absence of this grander network, these regions of the brain may still dutifully process signals streaming in from the skin, ears, and eyes.

That doesn't mean that anyone is home, so to speak, to perceive and become aware of those signals.

Owen and his team sought a more definitive measure. They needed a test that, if passed, would unequivocally demonstrate that a patient was conscious. They decided to give patients a verbal instruction and ask them to follow it. If these subjects heard the instruction, understood the command, and initiated an action in response, they were by definition conscious. It was setting a high bar indeed—requiring people with severe brain damage to be awake and alert, process verbal instructions, hold these instructions in working memory, and execute a response. Perhaps no one in a vegetative state would be capable of doing this. But in order to find out, Owen and his team had to devise a way for patients with no volitional control over the body, even down to the movements of their big toes and eyelids, to make a response on command.

The team's solution was rooted in brain maps and their reliable commonalities from one person to the next. As we have seen, generating visual imagery activates visual maps, and generating movement imagery activates movement maps. A conscious person trapped inside an unresponsive body might still be able to generate mental imagery, which Owen could then detect with functional MRI. So Owen and his team decided to place patients in an MRI machine and then ask them to imagine playing tennis. If they heard and complied, the resulting imagery could be detected as heightened activation in the patients' movement maps. They would also ask patients to visualize walking around the rooms of their house. If the patients heard and complied, the scientists would be able to detect this visual imagery as increased activity within their PPA place zones. The scientists attempted this test for the first time ever on Carol. They observed the activity they had hoped for in just the right brain areas and at just the right times. Carol had clearly heard them and followed their instructions. Despite her inert body, her brain was alert and aware—and now the experimenters knew it.

Owen and his colleagues worked for years to refine and improve this method. With it, they found that about one in five patients who appeared vegetative was still able to understand and mentally follow their verbal commands to create mental imagery related to movement or vision. Next, the team began to use their method to ask questions of these patients who could generate imagery on command. For example, they would instruct a patient to imagine playing tennis if the answer was yes and to imagine walking through his house if the answer was no. They began by asking questions with verifiable answers, such as "Is your father's name Thomas?" and "Do you have any brothers or sisters?" Accurate answers would confirm that the method could be used to communicate with a patient. They could even ask practical questions as to whether the patient was in pain or wanted to watch particular shows on television. This represented a remarkable first step in giving patients like Carol the chance to communicate.

Despite its tremendous promise, this approach also has significant drawbacks. One of them is cost. State-of-the-art MRI machines cost millions of dollars. Simply conducting a scan costs hundreds of dollars per hour. Because of these challenges, Owen's team and others have worked on developing less expensive alternatives that can be conducted at a patient's bedside. These use electrodes placed unobtrusively on the scalp to listen in on electrical signals coming from the brain. This technique for measuring brain activity, called electroencephalography, or EEG, has been around for more than a century and is far cheaper and easier to administer than functional MRI.

A second limitation is that we don't know what it means if a patient fails to show clear evidence of following the instructions. Eighty percent of the patients tested fell into this category. Are they less conscious than the other 20 percent, or perhaps not conscious at all? We have no way of knowing. But we do have evidence that, at least sometimes, the 80 percent can include people who are conscious. We know that because of Juan, a young man in a vegetative state whom the team tested twice. They found no clear evidence of consciousness in any of their tests. Yet in the months after he was tested, Juan mounted a remarkable recovery. He began talking, moving, and even walking again. When Owen's team interviewed him after his recovery, he recounted vivid memories of the tests and the MRI machine and the faces of the scientists who were with him that day. Despite all outward evidence and everything that the team could see on their scans, Juan had actually been alert and aware of his surroundings. Clearly a failure to detect consciousness does not necessarily equate to an absence of consciousness. The families of these 80 percent could walk away feeling as uncertain about a loved one's mental status as they felt before the test. Worse, the results could cause them to assume that their loved one was probably not conscious. Without knowing how often the test gets it wrong, we can't know whether this assumption is warranted, and we may leave these families in a particularly difficult position.

Despite these challenges, the potential benefits of identifying and communicating with conscious individuals trapped within inert bodies are truly breathtaking. Will there come a day when every patient in a vegetative state is assessed for brain-based evidence of consciousness? Ultimately the future of the technology—whether it will be refined, revamped, abandoned, or widely deployed—will depend on the numbers. How often is it wrong, and at what cost? How expensive is the test? How practical and useful is the information that it provides? The answers to these questions will determine whether and how we might help people like Carol regain a voice.

MIND READING WITH BRAIN SCANS

Like the brain scans that made contact with Carol and other patients in a vegetative state, the most successful forms of mind reading to date inform us about *whether* or *which*. In other words, these methods allow us to use thoughts and mental activity to distinguish between two or a few possibilities. In Carol's case, the presence or absence of increased activity in movement and object maps at the appropriate times informed the scientists as to whether the patient apparently in a vegetative state was actually aware and alert.

Another example of this kind of mind reading capitalizes on how attention boosts neural activity in specific regions of brain maps. If people are shown superimposed pictures of a face and a house while they are scanned in an MRI machine, scientists can tell if they are paying attention to the face or the scene based on the activity in certain areas of their brain, including the FFA face zone and the PPA place zone. Recall that activity in the FFA goes up when you attend to a face, whereas activity in the PPA rises when you attend to a landmark such as a house. This information gives scientists the ability to monitor people's moment-to-moment allocation of attention during a brain scan, and they can even use this information to train people to better focus their attention.

Yet much of what one might hope to read from a brain can be unlocked only through fancier means. For an example, let's return to the scans performed on Carol. The goal of that undertaking was to read whether Carol was creating imagery on command. But could the scientists have read more from Carol's brain activity? If she was imagining playing tennis, could they have deciphered exactly when she imagined swinging her racket and whether it was a backhand or forehand, a drop shot or volley?

By all rights, deciphering which tennis stroke Carol was imagining should be harder than determining whether she was imagining at all. In fact, until the past decade or so it was impossible. Why? Because brain maps, like physical matter, get weird when you zoom in too close. Just as matter acts differently at the scale of an atom than it does at the scale of a shoe or an asteroid, the layout of the brain looks different when you zoom in close to the fabric of the cortex.

When we zoom in close to a brain map and begin looking at clusters of neurons called columns, we find remarkable diversity and far more dimensions than previously imagined. Consider the V1 visual map, which is laid out according to the dimensions of your retinas. Retinal dimensions involve basic spatial dimensions: up-down and left-right. But vision has far more dimensions than these, and several are represented in the V1 map. For example, some cells in V1 respond to inputs from only the left eye or only the right eye. Others care about the orientation (vertical, horizontal, or oblique) of any edges that fall within their receptive fields. And some care about color. Inputs from a particular eye, edge orientation, and color: these are some of the key dimensions that the brain uses to tease form, surfaces, and depth from the light that enters our eyes. So, if you popped an electrode into the foveal region of someone's V1 map, any neuron you stumble upon would have a receptive field corresponding to the person's center of gaze. But interpreting what it *means* when that random cell fires depends on knowing which dimensions of vision that particular cell cares about.

Studies that have investigated the fine-grained organization of dimensions within brain maps find exquisite patterns. The illustrations in Figure 33 depict the same small area of V1. In the first frame, you see stripes that correspond to neuron preferences for inputs from the

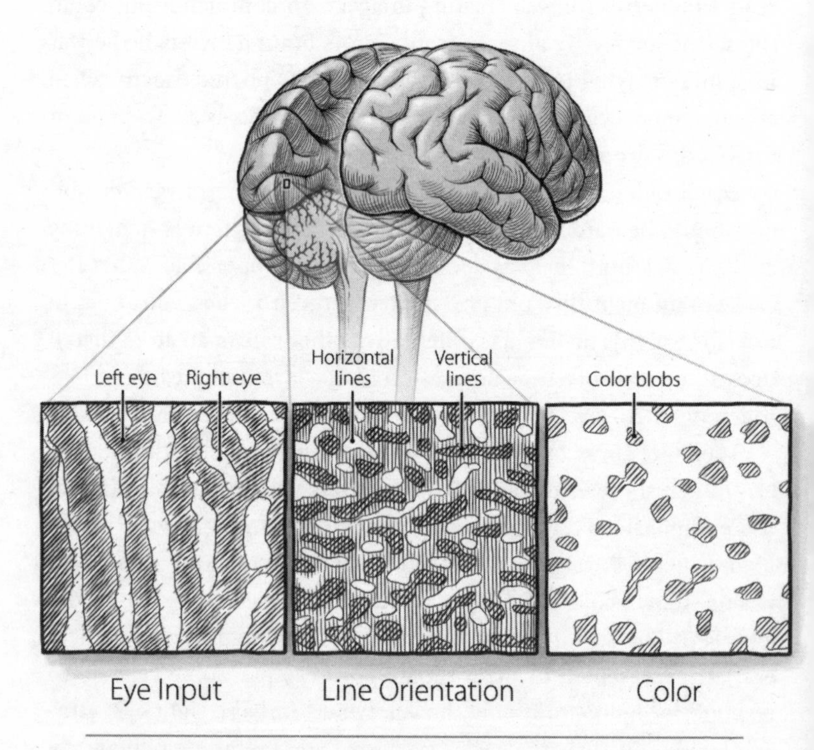

FIGURE 33. An illustration of the fine-grained, multidimensional organization within the V1 visual map, including dimensions for eye input, line orientation, and color. *Paul Kim*

two eyes. Neurons in the dark regions prefer input from the left eye, while those in the light regions prefer inputs from the right eye. The second frame shows orientation preferences, with light regions preferring horizontal edges and dark regions preferring vertical edges. The shaded regions in the third frame, known as color blobs, are selective for colors. As you can see, these preferences overlap. For example, most color blobs land squarely in the center of eye-input-preference stripes, which means that the neurons within them respond best to information about specific colors coming in from a specific eye.

Extra dimensions are also woven into our other sensory maps. Although touch maps like S1 are roughly laid out according to the dimensions of your body surfaces, they actually represent many other dimensions of tactile information. These include different physical features of touch (pressure, flutter, and vibration) as well as pain, heat, and cold. These dimensions may be specific to a particular creature or a particular body part. For example, the S1 maps of creatures with whiskers have magnified zones that represent touch for each individual whisker. If you zoom in on one of these individual zones, you will find buried within it a wealth of information about whisker deflection, or how a whisker is pushed up or down, left or right, from its normal resting position when it comes in contact with objects out in front of the animal. In fact, within each individual whisker zone in the rat's S1 map, scientists found a tiny map, laid out radially like a pinwheel and organized according to the direction of whisker deflection.

The intricate microstructure of brain maps can be a bit mind-bending, not unlike thinking about subatomic puzzles such as quarks and electron locations. These microstructures could be the subject of a whole book, but for our purposes it suffices to know that the brain on a small scale (on the order of tenths of a millimeter, or about one-hundredth of an inch) is both elaborately patterned and frustratingly complex. These patterns are comparable in size to the lines that make up human fingerprints, and like fingerprints, patterns across individuals follow similar rules and yet each is unique.

From the perspective of mind reading, this uniqueness poses a problem, as do the other differences between your brain and mine.

As we have seen, the exact layouts of our brain maps differ. The cognitive strategies we use and whether we can, for example, generate mental imagery may differ. And although the microstructure of your brain maps follows the same rules and makes the same basic patterns as mine, the precise layouts are unique. So if a technology's goal is to access *all* of the information contained in your brain, based on the activity of your neurons, it is destined to fail unless it can figure out precisely what each of those neurons is representing specifically in *your* brain.

I will say it plainly: we will never develop a technology that gives us access to all of the information represented in another person's brain, not because there is so very much information to be found there but because it lies buried within countless neurons nestled in their idiosyncratic stripes, blobs, and pinwheels and interfacing with one another through idiosyncratic and often changing connections. In order to access *all* of the information represented by a brain at a given point in time, we would need to know what is happening within every neuron in that brain and what each of those neurons represents. Given the constraints of anatomy and physics, we will probably never have such total access to the human brain. Just as there are tradeoffs when it comes to designing a brain, there are tradeoffs when it comes to observing and measuring what happens within it. As a consequence, any mind-reading technology must make an effort at reading minds *well enough*.

There are three specific challenges at the heart of these tradeoffs: how clearly we can listen in on individual neurons, how widely we can capture information from neurons across the entire brain, and how easily we can apply a technology across uniquely patterned brains. Consider the first of these challenges: getting the clearest possible signal from neurons in the brain. The closer we get to knowing the moment-to-moment activity of individual neurons, the better we can tease out information from their activity. Thanks to brain maps, we can tell a lot about what neurons probably do based on where they are located in the brain. Still, intricate microstructures like the stripe and pinwheel patterns ensure that even neighboring neurons can represent slightly different things. The only way to measure the activity of individual neurons is to stick electrodes directly on or inside a brain. Either way, this involves cutting open the skull and physically mucking around with a person's brain. That's a huge drawback. Another big drawback is that each electrode can listen in on only a handful of neurons. Although it is easy enough to put scads of electrodes in a brain, there are practical limits to how many neurons in how many parts of the brain these electrodes can reach without damaging the brain and killing the patient.

A second challenge lies in collecting information about activity happening throughout the brain. Recall how meaning is represented in many parts of the brain all at once. As such, there are limits to how useful information from any single neuron or even any single area of the brain can be. For the moment, functional MRI is the best technology we have to simultaneously monitor the activity of all neurons across the entire human brain. As an added bonus, functional MRI is safe and leaves a person's skull intact. But it will never provide a clear signal of what individual neurons are doing. The best it can do is give us an average of what groups of hundreds of thousands of neighboring neurons are doing at different places in the brain. That is powerful information to have, but it is a muddier signal than we could get if we listened using electrodes placed inside the brain. Moreover, the expense of MRI machines limits how widely mind-reading techniques based on functional MRI might be rolled out to the general public. These machines are also large and immobile, which makes them useless if the goal is to help people carry on with their daily lives.

The third challenge stems from the uniqueness of each brain. A mind-reading technology that is developed to beautifully read my mind based on the detailed microstructure of my brain maps may prove useless at reading yours. No technology can be rolled out for mainstream use if a team of scientists has to puzzle over how to tweak it for each new user. However, this challenge now has a solution: artificial intelligence, or specifically, a branch of artificial intelligence called machine learning. Here's the idea behind machine learning. Rather than teach a computer program a lot of facts to make it smart, we can give it the capacity to learn for itself, and then expose it to vast amounts of data. By learning from many examples and trying and failing exhaustively, the program can find the best solution for a problem or identify the crucial tidbits from a multidimensional tangle of information. These programs have become a powerful and pervasive force behind many of the technologies we rely on, from internet search engines to voice recognition technology. They excel at finding the key patterns buried within a mountain of data.

Machine learning programs are voracious omnivores. They will devour anything, in whatever quantities we throw at them. They can learn to predict future weather patterns based on current radar, to forecast a new product's popularity based on sales of previous products, or to detect early signs of disease in medical scans. The more relevant the data we train them on, the better they will be at detecting and predicting events when we show them something new, like a scan from a new patient who does not yet have a diagnosis. Eventually, the program's ability to detect or predict may outstrip that of the humans who dreamt it up. And *that* is when the program becomes truly useful. We can then turn to the program to make a diagnosis or tell us whether we should take an umbrella along when we leave the house.

Machine learning is a particular boon to mind reading because it overcomes a major obstacle: learning the specifics of how any individual brain represents stuff. It can thus save time and make it possible to tailor a mind-reading technology to a new brain without employing a costly team of scientists. A machine learning program trained to identify brain states based on brain measurements is called a decoder. Whether scientists take brain measurements with electrodes or with MRI machines, these measurements invariably produce numbers . . . lots and lots of numbers. Some reflect where a measurement came from in the brain. Others reflect the firing rates of neurons or the changes in blood flow that accompany changes in neural activity. The point is, decoders learn from these massive sets of numbers. They don't know or care what the numbers mean: they might describe forest growth or hot dog sales – or brain activity. A decoder's job is simply to find the useful patterns amid the data.

The fastest way to teach a decoder is with the help of the brain being read. This process begins when the brain's "owner" engages in key activities, such as making a certain movement or visualizing something on command. A decoder can then analyze the activity that took place in the brain during each key activity. In essence, by carrying out these activities on command, the brain is training the decoder how to go about reading it. That bizarre fact has a couple of upshots. One is that mind reading is not instantaneous, even with the help of a decoder; learning always takes time or, in the case of machine learning, examples. The other upshot is that decoding is most successful with the cooperation and patience of the brain that's being read. This matters because it makes it harder to read the mind of someone against their wishes.

If machine learning has more or less solved the third challenge to mind reading, what solutions exist for the first and second? Specifically, how do we balance the conflicting challenges of listening clearly to individual neurons while also listening simultaneously to many of them located all over the brain? As it turns out, machine learning is such a powerful tool that we can achieve *good enough* through a variety of compromises. One happy medium is to use patterns of activity gathered with functional MRI. Each measurement with functional MRI averages across the activity of hundreds of thousands of neurons. But if you feed a decoder enough examples of these patterns when the participant did either X or Y, the program detects the subtle relationships between the countless brain measurements from these two states, which allows the decoder to discriminate between them.

With these methods in hand, scientists have been using functional MRI to read minds for more than a decade. Based on the patterns of activity in a person's functional MRI scan, scientists can tell with reasonable accuracy what sort of thing, from a limited set of possibilities, that person was looking at, imagining, attending to, recollecting, or trying to hold in working memory at a given point in time. One study even focused on the contents of dreams; based on brain activity in the visual cortex during sleep, scientists could predict, with about 60 percent accuracy, whether a person was dreaming about, say, a person, a street, a car, or another type of object.

This decoding technology isn't limited to extracting seen or visualized images from the brain. Anything a person senses, imagines, or thinks about could be subject to decoding. One study involved playing audio clips of spoken sentences to people in an MRI machine. A decoder trained on their brain-activity patterns was then able to decode which speech sound the subject was hearing at any point in time, based on activity in the A1 sound-frequency map and other auditory regions in the temporal cortex. By training a new decoder differently on the same scans, the scientists could now decode which *speaker* the subject was hearing. Other studies have proved able to decode whether a participant was feeling pain, the meaning of the word that the participant was reading (out of a set of sixty words), which action a participant took or was about to take in a game, and whether they just won or lost a round. Those are just a few examples of what has become a widespread and popular approach to brain scan experiments.

Other studies have tried to actually reconstruct, based on brain activity alone, what a person in an MRI machine was seeing. This is a far bigger challenge than decoding which of a handful of states the brain is in. Instead of two or ten or even sixty possibilities to choose among, there are virtually infinite possibilities. A couple of ambitious studies tackled this challenge by combining machine learning with tons of information that neuroscientists have collected through experimentation about the properties of neurons in the V1 visual map. After the programs were trained on patterns of brain activity in a subject's V1 visual map, the programs attempted to reconstruct other paintings or shapes the subject saw, based on their brain activity alone. Given the challenges, the results were quite impressive, but they still fell far short of precisely re-creating what the subjects saw.

Reconstructing what a person is looking at based on their brain activity is scientifically interesting, but it does not have obvious practical utility. A far cheaper and easier way to know what a person is looking at is to turn your head and follow their gaze. Yet there are plenty of ways in which mind reading can have obvious, real-world utility. Perhaps the most high profile of these is the use of brain activity to determine whether a person is telling the truth. Using functional MRI, many studies have found differences in brain activity related to deception and truth-telling. These differences tend to be subtle and variable, but they are in some cases sufficient to permit reasonable detection of lies that subjects deliver while being scanned. Companies have emerged with the aim of using functional MRI to detect deception in romantic partners, employees, and even people on trial for committing a crime.

But there are serious concerns about the real-world value of brain-based lie-detection technologies. First, these methods are based on experiments done with young people, often college students, who are instructed to either lie or tell the truth while they are scanned. Of course, there are bound to be differences between how college students lie about inconsequential things *when they are instructed to do so* and how a criminal would lie about a high-stakes event when he or she is told to tell the truth. Lying when someone tells you to lie is hardly a lie at all. Second, criminals will probably deliver lies that they have practiced and repeated before, whereas the college students in the experiments are delivering lies that are new to them. A well-rehearsed lie is probably harder to detect than a spontaneous one. Another concern is that we have no way of evaluating how often this technology gets it wrong in real-world, high-stakes scenarios – either by failing to detect actual deception or, worse, by falsely detecting deception when a person is telling the truth. Finally, scientists have already shown that these tests can be manipulated. For example, associating imperceptible movements of your fingers or toes with questions to which you answer truthfully can disrupt scientists' ability to detect when you are actually lying.

If lie detection by functional MRI can be outmaneuvered by a deceptive subject, it is likely to go the way of the polygraph. It is not enough for a lie detection system to work on most people in most situations. To be used in high-stakes contexts like courtrooms, such a system would need to be extremely accurate, reliable, and resistant to manipulation. Although several court rulings have already deemed evidence from such methods inadmissible, it remains to be seen how future courtrooms around the globe will look upon the enterprise of reading deception from the brain.

The combination of functional MRI and machine learning has succeeded in lifting the veil on many aspects of human thought and perception. It will certainly expose more in the years to come. But its widespread utility is still severely limited by the expense, immobility, and scarcity of MRI machines. Beyond that, it works only on the brains of patient, compliant participants who are willing to keep still and follow commands. These limitations are welcome news to those who might worry about invasion of privacy. But that does not mean we have nothing to fear from efforts to read minds that are under development.

MIND READING AND WRITING FROM INSIDE THE BRAIN

Although we have techniques like functional MRI for eavesdropping on the brain without penetrating the skull, their signals will always be more corrupted and less precise than what you could record at the center of the action, where the actual thinking is taking place. So if a technology's goal is to read out clear details about what a mind is thinking, sensing, or trying to do, it should collect data from within the brain, if possible.

For an illustration of the benefits of eavesdropping on a brain from within, consider scientists' success at reconstructing the specific human faces that a subject is seeing at a given time based on brain activity alone. In this case, the subject was a monkey with electrodes implanted in the face zones of its visual object maps. The neural activity, recorded by the electrodes and run through scientists' models and a decoder, permitted reconstruction of almost the exact faces that the monkey had been viewing. See the results for yourself in Figure 34; the faces on the left are those that the monkey saw. Those on the right were reconstructed, based on the animal's brain activity alone.

Although reconstructing viewed images based on brain activity is impressive and informative to scientists, it is not particularly useful in a practical sense. That stands in stark contrast to the reconstruction of intended motor actions based on the activity of neurons in the motor cortex. Improvements in robotics have made movable prosthetic limbs possible. Now scientists are advancing technology that decodes motor activity from the brains of paralyzed people and uses

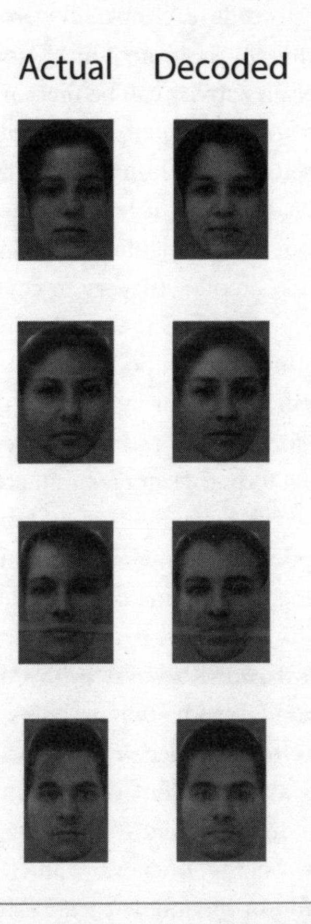

FIGURE 34. A demonstration of mind reading in monkeys. The images on the left were faces actually viewed by the monkey, whereas those on the right were decoded entirely from the monkey's brain activity. *From* Cell, *vol. 169, no. 6. Copyright © 2017 by Elsevier Inc.*

it to control assistive devices. This is the best possible application of mind reading: to empower the mind being read.

One of those minds belongs to Dennis Degray. He has been paralyzed from the neck down ever since he took out the trash one rainy morning, slipped, and fell. His paralysis left him unable to do many basic things for himself, from feeding and changing himself to holding a book or writing a letter. But thanks to an experimental procedure, Degray has the ability to use and control an iPad with his mind. Through a surgical procedure, electrodes were implanted directly in the hand region of his M1 movement map. Now, with the electrodes in place, Degray's brain activity can be measured and then analyzed by a decoder, which sends instructions to Degray's tablet, making the cursor move. The setup allows him to move the cursor and, from a display on the screen, to select the letter key he wants to type.

Since the electrodes are implanted in the hand region of his motor cortex, Degray uses motor imagery to control the movement of the cursor. As he explains it, "The visualization I found most comfortable for me was imagining a pool ball on a table. With my hand on top of the pool ball, you roll away, it goes up. You pull back, it goes down. Left, right accordingly. Like the old tabletop video games." The decoder then learned to pair Degray's brain activity during this motor imagery with specific movements of the cursor across the screen, allowing the conversion of Degray's imagery into visible action.

Scientists working at the same company have found other ways to use brain implants and mind reading to empower paralyzed patients. One of them is Bill Kochevar, who was paralyzed from the neck down after a biking accident injured his spinal cord. Since then, he has been confined to a wheelchair and dependent on caregivers to feed him, clean him, and turn him every two hours. Kochevar volunteered to participate in a different study aimed at restoring mobility after paralysis. Like Degray, Kochevar had electrodes implanted in the right-hand region of his motor cortex. Months later, the scientists also put electrical stimulators into the muscles of his right hand, arm, and shoulder. When turned on, these stimulators could make his muscles contract and move his hand and arm.

After quite a bit of practice, Kochevar could use motor imagery to make his arm move as he wished. The decoder read out the signals from his brain and converted them into instructions for the stimulators in his hand and arm muscles. The stimulators then set to work making his specific muscles contract as instructed. The result was that Kochevar could use his hand to grasp things and use his arm to lift food to his lips. "It was amazing," he said, "because I thought about moving my arm and it did. I could move it in and out, up and down."

There were hurdles. The stimulators could not generate a strong enough contraction in his right shoulder to overcome gravity and lift the full weight of his arm. So the scientists put his arm in a mobile arm support, which is like a motorized sling. The sling could lift and lower his arm and, like the stimulators in his muscles, it was placed under his brain's control. The process required a special rig with the support device, so he could use his implant only at a designated table in the clinic. The decoder also needed time to learn what Kochevar's motor cortex was telling it. And Kochevar had to practice for months before he could control the system effectively. But he was heartened, even ebullient, about the progress. The very notion that a quadriplegic man would be able to grasp a spoon and feed himself mashed potatoes is nothing short of astonishing.

Just as brain implants permit mind reading, they also make mind writing possible. Whereas mind reading involves observing and decoding activity in the brain to get information out of the brain, mind writing involves placing something—a sensation, a thought, or a desire—into someone's brain. Of course, you do this indirectly all the time. When you talk or write to someone, you are changing that person's brain state to make it align with your own. Every time someone gives a speech or airs an advertisement, they are attempting to change your thoughts and affect your behavior. But persuasion by these means is generally voluntary. If I don't want to hear your message, I can cut you off, walk away, cover my ears, or close my eyes. Mind writing is potentially more intrusive and harder to escape. For example, if I deliver a message directly to your V1 visual map, then you can't help but see it. Closing your eyes makes no difference at all.

Most mind-writing technologies under development aim to empower people with impaired hearing or sight by restoring some of their lost sensory abilities. Cochlear implants, medically accepted and widely used, are one example. These implants deliver information about sound frequency to the auditory nerve of people who are deaf or hearing-impaired. These implants involve the targeted use of a few electrodes within the inner ear and do not require penetrating the skull. Although the experience they provide is not like natural hearing, people with these implants learn to use the information they provide.

By definition, any mind-writing technology must be invasive, as it needs to directly alter the activity of your neurons. The most promising method for mind writing is to whisper directly into the brain through electrodes either implanted in it or resting on its outer surface. Recall the attempts by Giles Brindley, a pioneer in writing to the human mind. He implanted radio receivers and electrodes on the surface of the V1 visual map in a blind patient, in the hope of giving her back her sense of sight. When he stimulated her visual cortex with the electrodes, the patient reported seeing phosphenes. Most of these phosphenes could be distinguished from one another and could be used to make the patient identify very basic shapes. That was back in 1967. Brindley's paper describing the results ended with a rosy forecast: "Our findings strongly suggest that it will be possible, by improving our prototype, to make a prosthesis that will permit blind patients not only to avoid obstacles when walking, but to read print or handwriting, perhaps at speeds comparable with those habitual among sighted people." It seemed that, surely, a commercially viable version of the device would be on the market in no time.

Needless to say, it was not. We remain without an approved and commercially available cortical visual prosthesis more than fifty years later. But this was not for lack of trying. Several groups have vied to make Brindley's prediction a reality. For more than three decades before his death in 2004, William Dobelle led a high-profile effort to develop and test the technology in humans. The team devised a method of connecting a small video camera, embedded in a pair of glasses, to a computer that controlled the implanted electrodes. The goal was to convert the key visual information captured by the camera into the patterns of electrode stimulation of the blind person's visual map. When current flowed through the electrodes, the blind person saw phosphenes. And, in theory, the user could glean location and shape information from the patterns of phosphenes. Jens Naumann, a blind man implanted with two of Dobelle's prosthetic devices in 2002, described his first experience using it: "I was not able to make out any shapes or sizes of objects when seeing them. I was using the camera as a visual cane rather than a device for assessing objects."

On his first days of trying out the device, Naumann slowly learned how to use this strange new visual information. He could locate a phone on a table by sight. He also briefly drove a Mustang convertible in an empty parking lot—an event that was videotaped and touted widely in the media. But Naumann admitted that he could have managed both tasks without sight. He could *hear* where the phone had been placed well before he could visually determine where it was by using phosphenes. As for the car, he simply went forward and then back—at a walking person's pace. He noted that if the members of the research team had simply spoken an instruction, he could have handled the car equally well.

For a time, he could use the device at home. Naumann had eight children, and he had never seen any of them. He described using his device at a family dinner: ". . . a few dots of light in the direction of each child was enough for me to know that we were ready to eat. I heard each one get in their chairs, and the backs of the chairs, once visible as dotted lines, now changed to reveal a glob of dots that moved. This is what my kids looked like—their faces only two or three phosphenes as they faced me. But it was better than nothing."

Overall, the practical utility of the device was limited. It gave Naumann a better sense of where things were around him without having to touch everything with his hand or a cane. But he could not tell *what* anything was. The phosphenes did not fuse into shapes or faces, as he had hoped they would. And within weeks of first using his implant, Naumann found that his twinkling phosphene stars were fading and dying away. One day, he turned on his device and saw only ten phosphenes. Then five. And then two. Whatever limited utility he got from his implants in the beginning was entirely gone.

Like many recipients of experimental cortical prosthetics, Naumann also suffered because of his devices. Many of Dobelle's patients experienced headaches and chronic infections around the implantation site. Some described pus oozing from the site more or less constantly for years. Many, including Naumann, experienced seizures — a constant risk when stimulating the brain. These difficulties demonstrate the practical challenges of mind reading and writing from within the brain. Implants with wires tend to cause infections over time. Electrodes can damage cells and blood vessels in the brain, triggering a healing response that encircles electrodes with scar tissue. This scar tissue insulates the electrodes, blocking them off from the neurons they were intended to influence or measure; this is probably why Naumann's device quickly stopped giving him phosphenes. Changes in context, such as changes in the patient's emotional state, can also cause the neurons to change how they fire, throwing off the decoder. To keep the decoder accurate, the system may have to be reset, or recalibrated, sometimes even several times in one sitting. And with time, electrodes can become corroded by the wet, salty environment of the brain. Over the course of several years, an implant will have fewer and fewer functioning electrodes, and each will exchange weaker and weaker signals with surrounding neurons.

Despite these challenges, brain implant technologies have advanced by leaps and bounds. Implants are much smaller and able to record from many more neurons at once than they could in decades past. Some can even operate wirelessly. In short, while opening up the skull and implanting something in the brain will always confer risk, the implants themselves will keep getting better and safer in the coming years. Ongoing research and development in visual cortical prosthetics might even improve upon the deeper problem: that phosphenes are not like lights in an old-fashioned theater marquee. They do not predictably coalesce to form lines or shapes, not to mention real-world object surfaces. One reason for this failure may be the many dimensions woven into the fabric of each person's visual maps. Devices of the future may offer better results through a highly personalized approach that integrates information about each patient's unique, multidimensional visual maps. Or they might improve by delivering visual information to the brain in creative new ways. For example, a new approach that involves dynamically tracing letters and shapes across the surface of the visual cortex with electrical stimulation has yielded promising results.

Visual cortical prosthetics may become widespread even if they deliver an experience that is quite different from natural vision—assuming, of course, that they are safe and reliable. Exactly how functional these prosthetics need to be for patients and doctors to embrace them will depend on the alternatives. Some challenges could be overcome by cheaper and safer means using our other senses—such as hearing where the telephone is placed or feeling obstacles with a cane. People have learned to assess their spatial environment with sound, using echolocation, or with touch, using devices that can be carried or embedded in clothing. Advances in speech synthesis, speech recognition, and self-driving cars will offer blind people more alternatives for independence and interaction. And, in doing so, they will raise the bar for what people should expect from a visual prosthetic device.

Although mind-writing technologies typically aim to restore lost senses, the mind writing of the future needn't be so conventional. Why stop at feeding visual information to visual maps or auditory information to auditory maps? Animals implanted with electrodes can learn to use information about senses they don't naturally possess. For example, electrodes were implanted in the whisker region of adult rats' S1 touch map, while a device that detects infrared light was attached to the top of each rat's head. Infrared light is invisible to the naked eye, both for rats and for humans. When the infrared sensor detected infrared light, the implant stimulated the animal's S1 touch map. Then the bionic rats learned that they could earn food rewards by indicating which of several infrared lights was illuminated.

At first, the animals mistook the electrical stimulations that signaled the presence of infrared light as strange and unexplained whisker sensations. They scratched at their faces with their tiny paws. But in a matter of weeks, they learned to use the signals from the electrodes in S1 to determine which light was illuminated and thus get their food reward. They began sweeping their heads and the attached infrared sensors back and forth to identify where the signal was strongest. Within a month, they were consistently sensing the invisible light and getting their rewards.

This experiment is an example of sensory augmentation: using technology to enhance a creature's ability to detect physical phenomena. But studies like these don't create a new sense; they simply translate a new type of information so that it can be processed by a creature's existing senses. That is pretty cool, to be sure. But it can already be done without penetrating the skull. Every time soldiers don infrared goggles to observe enemy movements at night, they are translating infrared light into visual information. Every day, people use devices like compasses and radar to augment what their senses can tell them. These devices are certainly safer and cheaper than neural implants. So long as they continue to meet people's needs, augmented abilities like feeling infrared light or seeing geomagnetic fields will remain the parlor tricks of lab rats.

FORTUNE-TELLING WITH THE BRAIN

So far, we have considered emerging technologies for mind reading and mind writing. These approaches aim to extract or influence the contents of a person's thoughts, respectively. But there is another way that brain-based data can be used: to characterize individuals. As human beings, we spend a good deal of our time collecting information about each other: what we are good at or bad at, how we tend to react in different situations, whether we are trustworthy. This is invaluable information because it allows each of us to predict what others will do, which in turn helps us make better decisions about who to interact with and how best to do it. But this information proves useful only because, though we all change moods, preferences, and habits over time, our individual traits tend to remain stable.

Brain-based technologies can take this process of characterizing others a step further by using brain activity or organization, rather than behavior, to predict what a person will do or need or struggle with. Although the layout of a person's brain maps offers clues about their perceptual acuity and ability to harness mental imagery, many of the greatest successes in characterizing brains have come by measuring the degree of communication between brain maps. Recall that vision, like all cognitive capacities, is supported by the joint action of many brain maps working together and at essentially the same time. For example, maps with dimensions based on the retina, like the V1 visual map, send information to areas like the FFA face zone and the PPA place zone. But the face and place zones also send information right back to these retina-based maps. What is true in the workplace is true in the brain: a team that coordinates and communicates performs better than one that doesn't.

To measure teamwork between brain areas, scientists often ask participants to lie silently in an MRI machine and rest while their brain activity is measured with functional MRI. The scans allow scientists to compute functional connectivity—essentially how heavily different parts of that person's brain are communicating with one another. In a nutshell, functional connectivity tells us about the underlying architecture of a particular brain by characterizing how well the various bits of brain talk with one another.

It turns out that this architecture is stable over time and surprisingly informative when it comes to characterizing what a particular brain can or will do. For example, after measuring the functional connectivity among visual areas in their subjects' brains, scientists spent several days training these individuals to perform a difficult visual task. They found that people who showed greater communication among visual brain areas before training ultimately learned to do the visual task faster and better than the other participants. This study and others like it show that it is already possible to predict how well a person will learn or perform on a variety of new tasks, from learning a new language to exerting sustained attention, based on brain scans alone. Advances in machine learning have propelled this type of brain-based prediction forward by ferreting out the most informative patterns buried within heaps of neural data.

The reach of brain-based prediction extends beyond mere talents or even learning potential. It permits some predictions about what type of mental illness a person is suffering from and whether particular treatments are likely to help them. It has been used to predict whether children will benefit from tutoring in math or reading, whether kids will try alcohol or drugs soon, and even whether adults will gain weight in the coming year. In many cases, these predictions are far from perfect, but they are better than guessing or flipping a coin—sometimes substantially so.

Some brain-based predictions don't even require an MRI machine. Brain data can be procured from electroencephalography, or EEG, and its handful of listening electrodes that are placed on the scalp. For example, simple EEG signals recorded from babies within days of their birth have been used to predict whether they will be identified as poor readers or dyslexic eight years later. Other EEG signals recorded from prisoners while they did a basic self-control activity allowed researchers to predict which of them would be rearrested within four years of their release.

Might we someday have mainstream brain-based screenings that are used to assess or predict our abilities and deficits? The answer to that question will depend on how reliably and accurately scientists can coax these predictions from a variety of human brains. The higher the stakes of the prediction, the higher the threshold a technology must meet before it can be used commercially or medically. If they get there, many of these brain-based prediction techniques have the potential to help people get needed treatment sooner and perhaps more effectively than they otherwise would. But we would do well to start asking ourselves how we want our society to handle and harness brain-based predictions. If, one day, we have a reliable and accurate way of predicting whether a prisoner is likely to commit new crimes after release, how should we decide whether to consider or ignore this information? If brain-based measures can reliably predict that an infant will go on to struggle with dyslexia and will probably not benefit substantially from typical reading interventions, would this lead parents and schools to devote more or fewer resources to helping that child learn to read?

At the same time, we need to think about who should control these technologies and how they should be used. Already, companies are racing to design accurate, practical, and affordable brain-computer interfaces. Some of these efforts are focused entirely on improving the lives of patients through medical applications. Others are moving forward with two parallel goals: to help patients and to bring technological convenience to the general public. A private company owned by the famous tech entrepreneur Elon Musk is pursuing those parallel goals. In 2019, Musk headlined a press conference to unveil the company's plans for developing neural implants comprised of ultra-thin electrodes. Musk and others from his company talked about using the technology to help people with brain damage, but they also clearly envisioned broader commercial applications. Musk explained that one goal was to "achieve a sort of symbiosis with artificial intelligence."

Musk is not the only familiar name in the race to develop technologies to interface directly with the brain. Google's parent company, Alphabet, co-owns a company that is developing implantable neural devices. Facebook also has its own group that is funding research in this sphere; its scientists have shown that they can decode activity recorded from the brain's surface when a person speaks and then transcribe it quickly and accurately as text on a computer. Facebook has also announced work on a device that reads neural signals from the scalp, presumably like EEG but built into a pair of glasses. As Facebook's CEO, Mark Zuckerberg, explained in a 2017 Facebook post, "We're working on a system that will let you type straight from your brain about 5x faster than you can type on your phone today. Eventually, we want to turn it into a wearable technology that can be manufactured at scale."

For all of the excitement and promise of new technologies, the corporate positioning that undergirds them warrants careful scrutiny. Even before brain-based measures are thrown into the mix, we already have ample reason to be concerned about personal privacy.

With the aid of machine learning programs and mountains of data about your purchasing and other online behavior, companies already know more about you than you may realize. Behavior is a product of your brain and body, and the clues that it offers machine learning programs about your present neural state and your future behavior are truly startling. With it, a major retail company could tell that a teenager was pregnant even before her parents knew—based solely on her buying behavior. Programs used by Facebook can identify the emotional states of their users based on their posts—information that can be harnessed to influence others' emotional states or target advertisements to people who are feeling particularly insecure. Scientists who used machine learning programs to analyze the timing of people's keystrokes while typing in Facebook Messenger and other applications on their mobile phones could predict whether the person typing had a tendency for depression with over 80 percent accuracy. Other work along the same lines has allowed scientists to identify people experiencing early-stage Parkinson's disease based on the timing of their keystrokes on a keyboard and people with Alzheimer's disease based on movements captured by the accelerometers built into their smartphones. In short, it is already possible for companies to know more about some facets of your health than you may even know yourself.

Placing our individual neural data within reach of corporations and advertisers would exponentially increase this threat to privacy. If keystrokes can divulge brain disorders, imagine what your brain activity can reveal about your health, your mood, your tendencies and vulnerabilities, and your future choices. Even if those brain signals are recorded with electrodes embedded within glasses or a hat, they will generate troves of data that can be used to train machine learning methods to ferret out virtually any type of information about a person's traits, tendencies, and conditions.

This threat is magnified if your brain data lands in the hands of tech behemoths like Facebook or Google, which may already have detailed information about your moment-to-moment activities online. The more training a machine learning program receives on signals from a particular brain, the better it can learn what specific signals from that person mean. If a company can link up your activities at certain moments (say, messaging with friends or reading advertisements) with signals from your brain at those moments, it has all the information it needs to train a machine learning program extensively on you and your brain. Given the amount of time most people spend online and interacting with devices, it would not take long for such a program to know you better than you know yourself. And any information it extracts about your health problems, mood swings, physiological rhythms, insecurities, and other vulnerabilities becomes a valuable commodity to those who to wish to sell you things or influence your beliefs and behavior. If anyone can purchase information about when and how you are most vulnerable to manipulation, there are implications not just for marketing but also for social discourse and democracy. To some degree, these problems are already with us, but placing our neural data in corporate hands would undoubtedly make them worse.

In 2019, the UK's scientific academy, the Royal Society, issued a report on the development of brain-based technologies. It opened with an urgent call to action, stating, "The unavoidable point for opinion formers, decision takers, policymakers, and the public is that they have the opportunity to shape the future of neural interface technologies. These technologies are being developed now. Investment is accelerating. The impacts will be profound—and if they are to be positive ones, society needs to be engaged early and often."

But what exactly *can* be done to affect how these technologies are developed and used? Quite a lot, in fact. Individual countries and international organizations like the United Nations have a long history of forming commissions and issuing laws and declarations to oversee and regulate sensitive issues at the intersection of science and society. We already have strict rules governing the sale and use of human organs, gene-editing technology, and chemical, biological, and nuclear weapons. We have laws in place to restrict disclosure of patients' private medical information and to protect the rights of human research participants. There is no reason why the collection and use of neural data could not be similarly protected, provided that there is a public desire and legislative motivation to make it happen. A working group of concerned neuroscientists, ethicists, and engineers has already come up with several practical recommendations that governments should be taking into consideration now, to anticipate and avoid the misuse of brain data. Ultimately, the response from governments will hinge on whether the public appreciates the significance and urgency of this problem. In the meantime, companies are not waiting for us to do this. They will keep privately developing these technologies with their bottom line in mind.

Whether they breach the skull or eavesdrop behind the ear, brain-based technologies are coming. Some are already here. They will get better. They will probably not be able to steal your every thought or thrust you into an unwelcome virtual reality. But they will be able to detect your emotions and predict at least some of your actions, abilities, and ailments. These technologies have the potential to empower the powerless, but they might also threaten our privacy and lessen our personal sovereignty. If we write them off as science fiction and blindly leave their development in corporate hands, we might all be in trouble.

Although this book's examination of brain-computer interfaces ends here, it is actually only the beginning of their story. There will undoubtedly be advances in the hardware and software that make these devices possible. Likewise, there are sure to be many advances in scientists' understanding of brain maps and codes, paired with improvements in what machine learning can coax from a brain signal. Certainly, there will be new challenges and difficult decisions ahead regarding whether brain data has a place in courtrooms, classrooms, chatrooms, or anywhere else. Get ready. For better or worse, the portals are open. Now the race is on to decide what we will make of them.

12

How Brain Maps Weigh Us Down and How We Rise Above Them

As human beings, we have the spectacular if frustrating habit of posing the same question time and again: *Why?* As a student, I was drawn to study the mind and brain for perhaps the very same reason that you were drawn to read this book. I wanted to know why people perceive the world the way they do. Brain maps offer real answers to those questions. But what can they reveal about the bigger picture —why you are the way you are and why you think the way you do?

The single best answer scientists have given us is evolution. We feel and act and think the way that we do because feeling and acting and thinking this way helped our ancestors survive and find mates and produce offspring who themselves could survive, mate, and have offspring. No one disputes that our early life experiences matter as well. But in science as in popular culture, we tend to talk about early life experiences in terms of deviations from an ideal. Provide a baby brain with proper care and expected inputs, like warmth and words and the sight of human faces, and it should grow to become the brain specified by its genes. Fail to provide these things and you risk harming its proper development. In short, evolution gives us an endpoint —an adult brain that thinks in ways that helped our ancestors survive—while early life experiences either support or derail each child's brain in developing toward that goal.

Yet brain maps suggest a different explanation for why we think the way we do. To see how, consider how they cope with radical departures from our ancestors' environments. As much as conditions have changed over the generations, the earth's gravitational force, called G, has remained constant. What happens if you take a creature whose ancestors all lived and died under the constant tug of G and raise that creature under a different gravitational force? Would its brain maps be remodeled for this entirely new environment? Scientists tested this by raising rat pups in gondolas hanging from centrifuges. Some centrifuges were not turned on, so that the rats in those attached gondolas grew up under normal gravitational forces. Other centrifuges were turned on and left constantly spinning. Their rotation pushed the rats in the spinning gondolas against the floor, resulting in a constant downward force on the rats twice as powerful as G (called 2G). The rats lived like this for three weeks, until the scientists took them out and studied the layout of the forepaw regions of their S1 touch maps. (See Figure 35.)

The animals raised under normal gravity had vast zones of the S1 forepaw map devoted to representing touch information from the hairless skin on the bottom surface of each of their tiny little toes. Only small flecks of the map represented touch sensations from the root of each toenail, indicating when the nail is pushed up or down. But the rats raised with a double dose of gravity pushing them against the floor had radically different forepaw maps. The hairless skin on the bottom of their toes had little representation in the map, whereas the toenails reigned supreme. The sketch in Figure 36 illustrates the rats' resulting brainscapes. Early life in double gravity created rats that felt more with their toenails than with the pads of their toes.

After those first three weeks, the rats raised in 2G were moved to an environment with normal gravity and remained there for more than a month. Even after they had lived longer in G than 2G, their forepaw maps continued to be dominated by zones representing toenails. Those first three weeks in the 2G environment had etched their brain maps in ways that later experience could not entirely undo.

The layout of the S1 touch map in both sets of pups would have been molded by both genetic and environmental influences. But

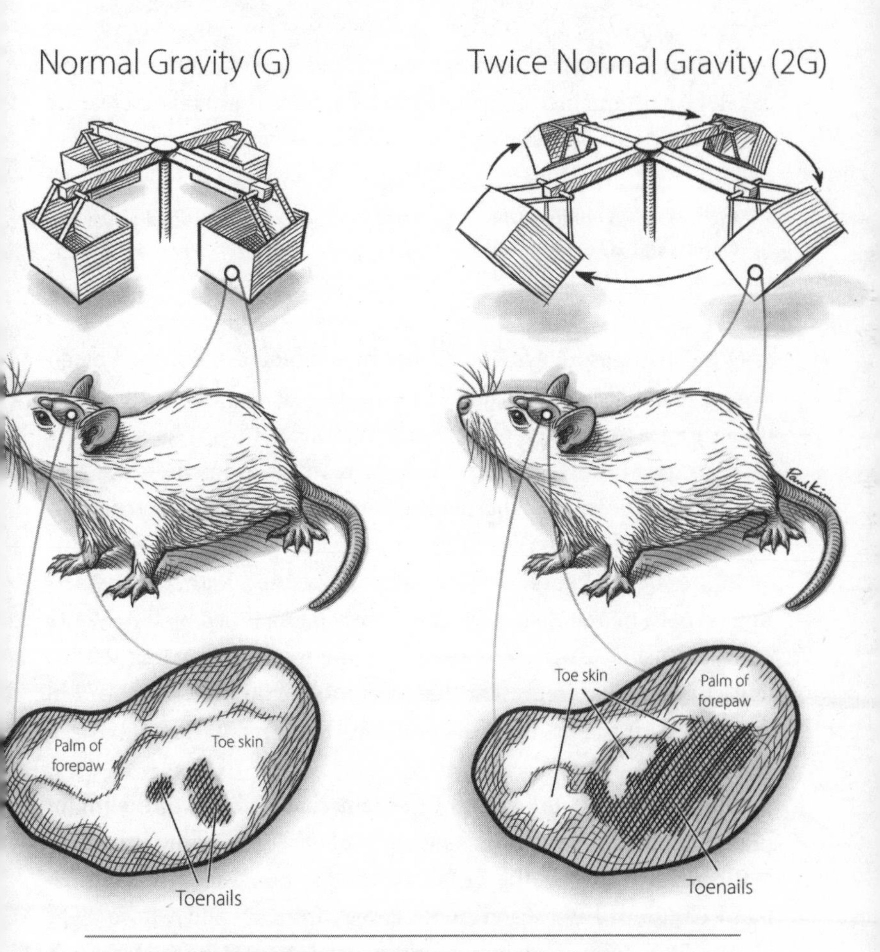

FIGURE 35. An illustration of the experimental design (top) and the impact of G versus 2G experience on the layout of the forepaw region of the rats' S1 touch maps (bottom). ***Paul Kim***

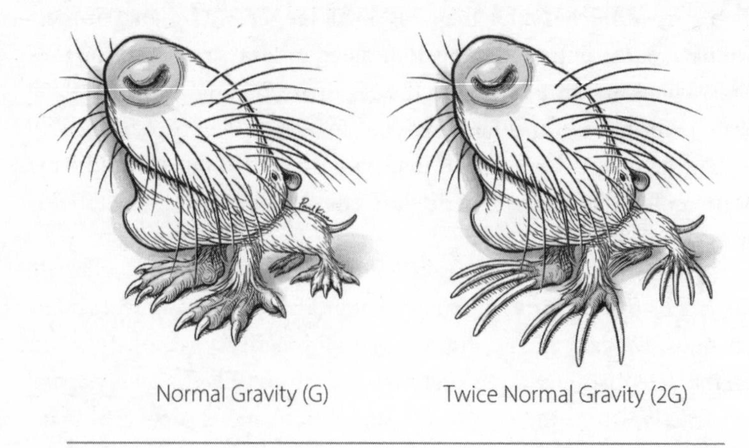

FIGURE 36. A comparison of the touch brainscape of rats raised under G (left) and 2G (right). *Paul Kim*

only the 2G pups allow us to pit hardwired blueprints shaped by an evolutionary past in G against the impact of early life experience under forces other than G. The fact that their S1 maps look entirely different from those of their normally reared cousins indicates that prenatal and neonatal experience can overwrite genetic instructions for the S1 map. It also illustrates how the impact of early life experience goes beyond good or bad, have or have not. Rather, every setting on each neural dial is simultaneously being tuned by the infant's earliest sensory experiences. In effect, the world is teaching the infant's brain how to represent that world to maximum effect. And in the process, it sculpts the child's or pup's brain maps, down to every last toenail.

Other lines of work support the remarkable ability of the infant brain to rapidly learn from its early inputs. These include the dramatic repurposing of the visual cortex for nonvisual functions in blind children. Another example comes from a child whose right hemisphere stopped growing early in fetal development. Although she was born with essentially half of a normal brain, she functions remarkably normally, aside from difficulty with moving the left side of her body. She experiences almost normal vision, even though she has only half of a typical V1 visual cortex. When scientists studied her V1 visual map using functional MRI brain scans, they discovered that the V1 map in her left hemisphere contains representations of both the left and right sides of visual space. In other words, her brain reworked its remaining half of the V1 map to represent twice the visual space it would normally cover.

Perhaps the most dramatic demonstration of learning early in life comes from a series of experiments carried out on newborn ferret pups. By damaging certain visual and auditory regions of their brains, scientists could redirect messages from cells that carry visual information from the eyes to the brain. Rather than deliver this information to visual parts of the brain, these cells began sending it to auditory areas, including the A1 sound-frequency map. Remarkably, a map of visual space developed within the auditory cortex of these animals. Moreover, the animals' behavior revealed that they experienced activity in this visual map as *seeing* rather than hearing. In other words, there is nothing preordained about the developing auditory cortex that makes it auditory, other than the fact that it normally gets its signals from the cochlea.

Such dramatic remapping simply does not happen in adult animals and humans. Although children born blind repurpose their visual cortex for nonvisual uses, those who become blind in adulthood cannot. A process of neural maturation and specialization takes place in infancy and childhood, crystallizing brain maps and neural representations. Through experience, adults can still refine their brain maps and representations; this ability underpins our capacity to continue learning throughout life. But adults cannot, say, convert an auditory map to a visual one, as the newborn ferrets did. This difference explains why people tend to recover much better if they suffer a stroke around the time of their birth than as an adult. The malleability of the infant brain allows it to rewire and remodel the intact portions of a wounded brain, enabling it to eke the most use out of the healthy tissue that remains.

Although the crystallized organization of mature brains is bad news in the context of recovering from brain damage, it is probably quite good news in the context of going about your everyday life. The entire reason that our brains are so useful to us is that they strike an ideal balance between memory, or stability, and learning, or change. Imagine if you had to wake up every morning and relearn how to brush your teeth and tie your shoes, not to mention newly master what every word means and how they combine to form sentences. Your brain's stability allows you to retain your knowledge and abilities from day to day and year to year. And yet if your brain was too stable, you would be unable to learn anything new or adapt to new events or circumstances.

So you see, the infant brain is a brain with the dial set to LEARN. Particularly in the case of humans, infants *do* very little but *learn* very much. And they don't just learn what we see them learning, like how to smile or say their first word. They learn about how and what to sense, perceive, and attend to. We are talking about a powerful, multifaceted kind of learning that involves sucking information of all kinds from the environment and using it to sculpt how information will be processed from here on out. We undergo a profound form of learning early in life, followed by a lifetime of moderate stability.

Given this profound early learning, it is important to recognize how children's experiences differ from those of adults. Even though children live in our homes, in many ways they do not inhabit our world. For instance, babies see the world through unfocused eyes and from different physical vantage points. When they are hungry, they cannot head to the kitchen to fix a snack. When they are cold or wet, they can't relieve their own discomfort. Although babies can be affected by financial distress, they know nothing of numbers, not to mention income shortfalls or debts. Their relevant world is one of space: *where* things are that they might see or grab. They have little need to know *when* or for *how long* things happen, beyond the value of recognizing that certain things tend to co-occur. Even as babies grow into toddlers, then preschoolers, and beyond, they are often unaware of the activities and concerns that are the hallmark of modern life as an adult. They have different priorities, notice different objects in their environment, fear different things, and seek out different rewards.

Some aspects of early experience are more or less universal to infants the world over. The similarities of our physical experiences as infants will drive similarities in our brain maps, above and beyond what is written in our genetic blueprints. And where those early life experiences are different from our experiences as adults, it will be the *infant's* experiences, and not the adult's, that dictates how the brain's basic territories and zones are laid out.

This idea might take a moment to sink in because it runs counter to the prevailing notion that childhood is merely a process of becoming an adult. We conceptualize childhood this way, both in popular culture and in science, and it colors how we interpret objective evidence. For example, in the past decade, scientists have developed safe and comfortable ways to collect functional MRI brain scans from children, including infants. These scans have revealed that, in its basic layout and function, the baby brain looks remarkably like the adult brain. Scientists have marveled at how precocious the infant brain must be to resemble the adult brain so early in development. But by all rights, we ought to recognize these findings for what they are: evidence that we adults go through life relying upon brains that resemble a baby's.

The notion that we muddle through adult life with brains that are in many ways built for a child raises an obvious question. If the adult brain fundamentally resembles the child brain, why are adults capable of so much more than infants and young children? The answer is that we learn and develop new abilities by building upon existing ones. In effect, we erect cognitive ladders to rise above our limited neural foundations.

In this book, you have already seen many examples of humans harnessing early-developing capacities in service of acquiring new ones. Visual perception provides the raw materials for visual working memory and mental imagery. Auditory perception lays the groundwork for speech perception, which in turn supports development of verbal working memory and the ability to read. Finger awareness supports early enumeration and calculation. Spatial processing promotes focused attention, then reasoning about number and time. In this way, we learn how to wring more abilities out of the same basic maps and neural architectures that we had as children.

We also manage to wring more functionality from our brains by outsourcing cognitive burdens. For example, written language and mathematical notation allow us to overcome our limited capacity for memory by preserving information in a physical format. We use machines to carry out most nontrivial mathematical calculations. We rely on calendars, timers, and clocks to help us keep track of time. In short, many of our adult capacities for complex thought and planning depend, at least in part, on our effective use of tools.

The vital concepts of time and quantity in particular pose a challenge for the human brain. Recall that our brains lack adequate time and number maps. As a result, we lean heavily on the aid of written notation and devices. For what we cannot outsource, we have struck upon a poetic solution. We employ an analogy, such as *Time is like distance in the space around us.* Our brains can flexibly align points in time with points in space (like *past is to the left and future is to the right,* or *past is behind me and the future lies ahead*), allowing us to use our spatial maps to represent and reason about time. We think about quantity using a similar analogy, that number is like space. These analogies enable us to think and talk about time and number with relative ease. The truly brilliant thing about using analogy in this way is that it overcomes a weakness with a strength. We are fantastic at thinking about the physical space around us. And so if time and number are like space, we can be instantly great at thinking about time or number, at least insofar as they are truly similar to space.

If you consider any of the spectacular accomplishments of the human mind, like sending humans into space and discovering what lies at the heart of an atom, you will see these tricks and tools in action. People's abilities to read, write, and perform mathematical operations leverage shape recognition, speech processing, and motor and spatial maps. They contemplate complex data by creating graphs that depict number, time, or other dimensions in terms of *space,* like distance in a bar graph or area in a pie chart. They use their sensory brain maps to imagine ways to overcome obstacles. They seek to understand complex phenomena by relating them to more familiar things, from *electricity is like flowing water* to *the brain is like a computer.* In the end, your ability to ponder the far reaches of the universe or the distant past is built atop a teetering tower of ladders. And beneath those ladders lie the basic neural foundations of sensation, action, and spatial awareness that were laid down in your childhood. In this way, your brainscapes undergird and fundamentally shape how you think, reason, comprehend, and imagine for a lifetime.

In many ways, your brainscapes and mine will be similarly molded by our similar experiences as babies and children. In others, differences in our genes, our fetal environments, and our early life experiences will forge differences in our neural topographies. In recent decades, there has been a growing awareness that fetal and early life experiences can have a powerful, long-term impact on each individual's health and well-being. Scientists and physicians have now uncovered many examples of these long-lasting effects on metabolism, immune function, and physiological reaction to stress. One comes from the Netherlands, where provinces occupied by the Nazis during World War II experienced mass starvation in the winter of 1944 to 1945. The famine was intense but not long-lived. Late spring brought liberation by the Allies, and the Dutch people enjoyed prosperity in the years that followed. But the children who had been conceived or born during that terrible winter bore an invisible scar. Their mothers' starvation shaped their fetal environment and triggered their developing metabolism to prepare for a lifetime of famine. As a result, these children would go on to have unusually high rates of metabolic disruptions linked to type 2 diabetes and obesity later in life.

Another example comes from newborn rat pups. When rats are born, they are blind, hairless, and unable to walk. Mother rats are kept busy nursing and licking their pups. Some mother rats lick their pups a lot in their first week of life; others, very little. Stressed mother rats tend to lick their pups less. Remarkably, this simple difference between being licked a little or a lot in the first week of life will determine the magnitude of each pup's lifelong physiological reaction to stress. Pups who are licked less grow up to have higher levels of stress hormones in their bodies. They are more likely to respond to situations with fear, to startle easily, and to be wary of new environments. In short, early experiences caused their brains and bodies to prepare for a life of uncertainty and danger. Similar mechanisms appear to be at work in human development. Children who experienced abuse, stresses associated with poverty, or social interactions affected by parental stress or depression tend to grow up to have higher levels of stress hormones and stronger reactions to potential threats. This strong stress response strains the heart and immune system and puts people at higher risk for physical and mental illness later in life.

These examples, and many more like them, demonstrate how experiences in utero and childhood can program our bodies and alter our physiology for a lifetime. This idea has become known as biological embedding. It helps scientists explain why adversity early in life leads to higher rates of illness in adulthood and, in doing so, magnifies the long-term toll of social inequality. Although biological embedding is typically used in the context of deleterious early experiences, its implications are wider reaching. Your early life—a time before you can even remember—was a period of dramatic learning from the environment, both for your body and your brain. Regardless of what that environment was, it burrowed into you, setting many of the dials for who you would become.

Biological embedding tells the story of the tissues of your body and brain and their biochemical and structural learning. As the pages of this book describe, there is a parallel story to tell about brain maps and the sensory, motor, spatial, and conceptual capacities they support. I will call this representational embedding: how early life experiences can shape neural organization and representation, affecting perception, action, and cognition for a lifetime.

The long-lasting impact of early brain organization, paired with our use of cognitive ladders later in life, explains why the infant brain has proved to be a powerful predictive tool. Already we have seen how brain signals can be surprisingly informative for predicting a person's future actions, abilities, and difficulties. In one of these examples, a neural signature recorded at a baby's scalp within days of birth could be used to predict whether the baby would struggle with reading eight years later. This EEG signal was recorded while the babies heard brief audio recordings of speech and nonspeech sounds. Although newborn babies will not begin learning to read or even recognize letters for several years to come, they already have experience hearing speech and other sounds in their fetal and neonatal environments. Years later, they will rely heavily on the link between speech sounds and written letters as they learn to read and write. And although reading in adulthood is typically a silent activity, their adult brains will harness hearing and speech-processing regions in the service of reading for the rest of their lives. That is why the processing of speech sounds in newborns can predict reading difficulties later in life. Newborns already have the foundations of auditory brain maps and speech representations upon which the cognitive ladders for reading will later be built.

The idea that early learning constrains later learning can be disheartening, even downright depressing. But there are substantial benefits to recognizing how adapting and learning in early life can shape everything that follows. It attests to the long-term societal benefits of investing in programs that support families and children. Biological embedding shows how stress or deprivation in infants' environments can raise the risk of physical and mental illness later in life. Likewise, representational embedding explains how isolated and unstimulating environments experienced early in life can curtail the development of the diverse maps and representations that support later-developing capacities like reading, calculating, and planning for the future. Together they reveal the prescience of paid parental leave, affordable early child care, universal preschool, and enriched primary education for all children. Given the long-lasting impact of childhood experience, protecting and supporting children is both a moral imperative and a wise investment for every society.

That said, we also need to rethink how we understand and talk about child development and early life experience. Childhood is a period of intense learning and preparation that affects us all. And how useful that preparation turns out to be will depend, for each of us, on what the future holds. All we can do is try to give children the enrichment and interaction that will help them develop diverse maps, because these neural foundations will give them the greatest ease and versatility for building cognitive ladders throughout life. We can also provide extra support for adults who lacked enrichment and interaction in childhood, to give them the time and tools they may need to adapt to new circumstances and build new ladders.

Regardless of our early life experiences, our genes, the money in our pockets, and the security of our surroundings, each of us lives and dies with a limited brain. Thankfully, our capabilities are far less constrained. Whenever the human brain has failed us, we have built more ladders. We overcame our limited memory capacity by developing a writing system. We overcame our deficient neural tools for numeracy by developing a formal system of mathematics. We used numbers to characterize and make predictions about the physical world and then tested those predictions with experiments. We used math and science to build technologies that can take us to the moon and edit our genetic code. We have invented not one, but many ways to peer into the living, thinking brain and watch it churn. Ladders and tools can take us where we need to go. Limited or not, we can get there, if we are willing to train our mental, physical, and cultural resources on what is coming next.

When I first set out to write this book, my goal was to share the beauty and explanatory power of brain maps. Although the brain operates on many levels and by many mechanisms all at once, brain maps in particular lend us a tangible roadmap for understanding how and why we think and feel and act the way we do. It was only later, while

I was in the process of committing words to the page, that I began to fully appreciate how far-reaching this topic truly is. Brainscapes are the deepest foundations of who and what you are. How they arise, adapt, and represent are integral to human health and the promise or peril of emerging technologies. Their uniqueness shapes our individuality. And their universality, across individuals, cultures, and even species, reveals how much we are the same.

Brainscapes also lay bare the remarkable relationship between our minds and the wider world. The physical environment, acting on your physical body, literally shaped how your brainscapes developed in childhood. And then those brainscapes became the conduits through which you would perceive the wider world forevermore. We are not islands unto ourselves—minicomputers propped up on two legs or fleshy sacks of DNA and water. We are more than the product of our mothers and fathers and all of the mothers and fathers who came before them. We are built by the sun's light and warmth, by the resonating air around us, and by our planet's constant pull. Above all, I am struck by the significance of this simple idea: our brainscapes embody the world's place within each of us.

In the process of bringing this book from idea to realization, I have witnessed firsthand how brain maps made that transformation possible. Every word in this book began as speech inside my head. I spoke these words to you with imagined speech, using auditory and movement maps to select my words and commit them to writing. I thought about you and imagined who you might be and what you might already know. With more internal speech, I read what I wrote and imagined what it might mean to you—how well it might align our brain states so that the meaning I hoped to convey would come across, rather than some other meaning or no meaning at all.

Any book worth reading should change how the reader experiences the world, even if only by a little. I hope you have found that true for this book. And I hope you will take a moment, at the close of our journey, to marvel with me at the tiny miracle we carried out together. I spoke to you in my head. Perhaps you listened to me in yours. I wrote by imagining what you might know and think. And you read by imagining what I meant. We communicated by inhabiting each other. We shared a multitude of brain states over time and space. It is entirely normal and mundane that we should do such a thing. But that doesn't make it any less astonishing.

Acknowledgments

Writing this book has been an exercise in gratitude and humility. Without the support, guidance, and forgiveness of many people, it would never have been possible or even imaginable.

First, I would like to thank colleagues and friends who offered advice and encouragement or gave feedback on portions of the manuscript: Paul Bloom, Sandra Blakeslee, Josh Witten, Sandra Aamodt, Talia Konkle, Cass Sunstein, Iva Freeman, Maya Rosen, Arvin Dang, Joan Luby, Deanna Barch, Chad Sylvester, Michael Graziano, Adrian Owen, and Joe LeDoux. Thank you to Chris White in particular, who encouraged me from the very beginning and shared his unflagging advice and support ever since. I am also grateful for the patience and support of my research mentors and role models during this process, including Nancy Kanwisher, Noa Ofen, Lori Markson, Chad Sylvester, Kirsten Gilbert, Joan Luby, and Deanna Barch.

My heartfelt thanks also go out to the institutions and resources that made this book possible. Thank you to the Alfred P. Sloan Foundation for supporting this book with a generous grant through its Public Understanding of Science, Technology, and Economics program and for its vital work in supporting and sharing scientific discovery. Thank you as well to the Washington University in St. Louis libraries, the MIT libraries, the UCLA libraries, MOBIUS, the Cambridge Public Library, the Municipal Library Consortium of St. Louis County, the St. Louis County libraries, and to open-stack libraries around the world that make knowledge freely available to all.

I want to thank Paul Kim for creating the beautiful original illustrations that brought this book to life. Thank you to my agent, Katinka Matson, for believing in this book and finding it a home. Thank you to Michael Healey for practical advice and Russell Weinberger for taking *Brainscapes* global. Thank you to Susanna Brougham for thoughtful and invaluable edits and to Lisa Glover for putting it all together. Thank you to the book's wonderful editor, Alexander Littlefield, for his dedication to this project from beginning to end. I am truly grateful, both to Alex and to Olivia Bartz, for showing me how to make the book clearer and more compelling.

To my family, thank you for forgiving missed vacations and work weekends, for helping with meals or child care, for your patience and your love. Thank you to Meena and Naya for their enthusiasm and sage book-writing advice. And thank you especially to Rajni Dang, to Terri Frye, to Sally Frye Schwarzlose, and to my husband, Sabin, for making this book possible by being there and helping me over and over again.

Finally, I want to say thank you to a few special people who will never hold this book but gave me the tools to write it. Thank you to Les Plesko for teaching me how to write. Thank you to my father, Richard Schwarzlose, for showing me how to research deeply and think expansively. And thank you to my mother, Sally Frye Schwarzlose, for teaching me how to read, to empathize, and to wonder. Above all, this book was inspired by my mother's sense of wonder and her uncanny ability to see the beauty within and around us.

Notes

1. AN ATLAS OF YOU: WHAT IS A BRAIN MAP?

page

7 *Many of these cases:* Saiichi Mishima, *The History of Ophthalmology in Japan* (Belgium: J. P. Wayenborgh, 2004); Mitchell Glickstein & David Whitteridge, "Tatsuji Inouye and the Mapping of the Visual Fields on the Human Cerebral Cortex," *Trends in Neurosciences* 10 (1987): 350–53; Danny H.-Kauffmann Jokl & Fusako Hiyama, "Tatsuji Inouye—Topographer of the Visual Cortex, Exemplar of the Germany-Japan Ophthalmic Legacy of the Meiji Era," *Neuro-Ophthalmology* 31 (2007): 33–43; Charles G. Gross, *Brain, Vision, Memory: Tales in the History of Neuroscience* (Cambridge, MA: MIT Press, 1998); Inouye Tatsuji, *Die Sehstörungen bei Schussverletzungen der kortikalen Sehsphäre: Nach Beobachtungen an Verwundeten der letzten japanischen Kriege* (Leipzig, Germany: W. Engelmann, 1909).

14 *the Swedish neuropathologist:* Salomon Henschen, "On the Visual Path and Centre," *Brain* 16 (1893): 170–80.

19 *Washington and his spies:* John A. Nagy, *George Washington's Secret Spy War: The Making of America's First Spymaster* (New York: St. Martin's Press, 2016).

20 *a group of vision scientists:* Roger Tootell et al., "Functional Anatomy of Macaque Striate Cortex: II. Retinotopic Organization," *Journal of Neuroscience* 8 (1988): 1531–68.

2. THE TYRANNY OF NUMBERS: WHY BRAIN MAPS EXIST

26 *as the tyranny of numbers:* Jon Gertner, *The Idea Factory: Bell Labs and the Great Age of American Innovation* (New York: Penguin, 2012); Arnold Thackaray, David C. Brock, & Rachel Jones, *Moore's Law: The Life of Gordon Moore, Silicon Valley's Quiet Revolutionary* (New York: Basic Books, 2015); Michael S. Malone, *The Intel Trinity: How Robert Noyce, Gordon Moore, and Andy Grove Built the World's Most Important Company* (New York: HarperCollins, 2014).

27 *using calls with a dialect:* Olga Filatova et al., "Cultural Evolution of Killer Whale Calls: Background, Mechanisms, and Consequences," *Behaviour* 152 (2015): 2001–38.

there's Clark's nutcracker: John M. Pearce, *Animal Learning and Cognition: An Introduction,* 3rd ed. (East Sussex, UK: Psychology Press, 2008).

The brains of these animals: Gerhard Roth & Ursula Dicke, "Evolution of the Brain and Intelligence," *Trends in Cognitive Sciences* 9 (2005): 250–57.

28 *neuron size and density:* Suzana Herculano-Houzel, "The Remarkable, Yet Not Extraordinary, Human Brain as a Scaled-up Primate Brain and Its Associated Cost," *Proceedings of the National Academy of Sciences of the United States of America* 109 (2012): 10661–68.

29 *This one chunk of tissue:* Christopher Kuzawa et al., "Metabolic Costs and Evolutionary Implications of Human Brain Development," *Proceedings of the National Academy of Sciences of the United States of America* 111 (2014): 13010–15.

30 *evolution has added neurons:* Kechen Zhang & Terrence Sejnowski, "A Universal Scaling Law Between Gray Matter and White Matter of Cerebral Cortex," *Proceedings of the National Academy of Sciences of the United States of America* 97 (2000): 5621–26.

more than 20 kilometers: Mark Nelson & James Bower, "Brain Maps and Parallel Computers," *Trends in Neurosciences* 13 (1990): 403–8.

34 *Mariotte was abbot:* Andrzej Grzybowski & Pinar Aydin, "Edme Mariotte (1620–1684): Pioneer of Neurophysiology," *Survey of Ophthalmology* 52 (2007): 443–51.

36 *one small white paper circle:* Edme Mariotte, "Nouvelle découverte touchant la veüe," in *Oeuvres de M. Mariotte* (The Hague: Jean Neaulme, 1740), 495–34.

39 *actual slices of a human brain:* Daniel L. Adams et al., "Complete Pattern of Ocular Dominance Columns in Human Primary Visual Cortex," *Journal of Neuroscience* 27 (2007): 10391-403

40 *one of the blind-spot regions:* Hidehiko Komatsu et al., "Neural Responses in the Retinotopic Representation of the Blind Spot in the Macaque V1 to Stimuli for Perceptual Filling-In," *Journal of Neuroscience* 20 (2000): 9310–19.

41 *the V1 maps in living humans:* Ming Meng et al., "Filling-in of Visual Phantoms in the Human Brain," *Nature Neuroscience* 8 (2005): 1248–54.

42 *functional MRI studies:* Hiroshi Ban et al., "Topographic Representation of an Occluded Object and the Effects of Spatiotemporal Context in Human Early Visual Areas," *Journal of Neuroscience* 33 (2013): 16992–7007; Gennady Erlikhman & Gideon P. Caplovitz, "Decoding Information About Dynamically Occluded Objects in Visual Cortex," *NeuroImage* 146 (2017): 778–88.

3. HOW BRAIN MAPS DETERMINE WHAT WE SEE AND FEEL

48 *hundred times more space:* Robert Duncan & Geoffrey Boynton, "Cortical Magnification Within Human Primary Visual Cortex Correlates with Acuity Thresholds," *Neuron* 34 (2003): 659–71.

to be thirteen times larger: Duncan & Boynton, "Cortical Magnification."

50 *people's V1 brain maps:* Duncan & Boynton, "Cortical Magnification."

51 *Brindley, a physician:* Giles S. Brindley & W. S. Lewin, "The Sensations Produced by Electrical Stimulation of the Visual Cortex," *Journal of Physiology* 196 (1968): 479–93.

54 *about the size of objects:* Leslie R. Newsome, "Visual Angle and Apparent Size of Objects in Peripheral Vision," *Perception & Psychophysics* 12 (1972): 300–304.

Penfield took his scalpel: Wilder Penfield, *No Man Alone: A Neurosurgeon's Life* (Boston: Little, Brown, 1977).

56 *Penfield and his colleagues:* Wilder Penfield & Edwin Boldrey, "So-

matic Motor and Sensory Representation in the Cerebral Cortex of Man as Studied by Electrical Stimulation," *Brain* 60 (1937): 389–443.

59 *people's tactile acuity:* Robert Duncan and Geoffrey Boynton, "Tactile Hyperacuity Thresholds Correlate with Finger Maps in Primary Somatosensory Cortex (S1)," *Cerebral Cortex* 17 (2007): 2878–91.

size based on touch: Barry G. Green, "The Perception of Distance and Location for Dual Tactile Pressures," *Perception & Psychophysics* 31 (1982): 315–23; Roger W. Cholewiak, "The Perception of Tactile Distance: Influences of Body Site, Space, and Time," *Perception* 28 (1999): 851–75.

61 *Adrian was equally comfortable:* Alan Hodgkin, "Edgar Douglas Adrian, Baron Adrian of Cambridge," *Biographical Memoirs of Fellows of the Royal Society* 25 (1979): 1–73.

"the most glorious clutter": J. K. Bradley & E. M. Tansey, "The Coming of the Electronic Age to the Cambridge Physiological Laboratory: E. D. Adrian's Valve Amplifier in 1921," *Notes and Records of the Royal Society of London* 50 (1996): 217–28.

62 *"a rushing noise":* Edgar D. Adrian, "Afferent Areas in the Brains of Ungulates," *Brain* 66 (1943): 89–103.

62 *Adrian excused his assistant:* Edgar D. Adrian, "The Somatic Receiving Area in the Brain of the Shetland Pony," *Brain* 69 (1946): 1–8; Hodgkin, "Edgar Douglas Adrian."

"The area is divided": Adrian, "The Somatic Receiving Area."

animals bring their nostrils: Edgar D. Adrian, *The Physical Background of Perception* (Oxford: Clarendon Press, 1946); Adrian, "Afferent Areas"; Adrian, "The Somatic Receiving Area."

63 *whisker endings, was enlarged:* Edgar D. Adrian, "Afferent Discharges to the Cerebral Cortex from Peripheral Sense Organs," *Journal of Physiology* 100 (1941): 159–91; Adrian, "The Somatic Receiving Area."

"The pig's snout is": Adrian, *The Physical Background.*

hidden portion of the map: Sandra L. Craner & Richard H. Ray, "Somatosensory Cortex of the Neonatal Pig: I. Topographic Organization of the Primary Somatosensory Cortex (SI)," *Journal of Comparative Neurology* 306 (1991): 24–38.

64 *part of the head or face:* Adrian, *The Physical Background.*

"the explanation is probably": Adrian, *The Physical Background.*

65 *representation of those teeth:* Kenneth C. Catania & Michael S. Remple, "Somatosensory Cortex Dominated by the Representation of Teeth in the Naked Mole-Rat Brain," *Proceedings of the National Academy of Sciences of the United States of America* 99 (2002): 5692–97.

66 *the creature's tiny nose:* Kenneth C. Catania & Jon H. Kaas, "The Un-

usual Nose and Brain of the Star-Nosed Mole," *BioScience* 46 (1996): 578–86; Kenneth C. Catania & Fiona E. Remple, "Tactile Foveation in the Star-Nosed Mole," *Brain, Behavior and Evolution* 63 (2004): 1–12.

67 *how impressive the pig's snout:* Adrian, "Afferent Areas."

68 *the rat's S1 touch map:* John K. Chapin & Chia-Sheng Lin, "Mapping the Body Representation in the SI Cortex of Anesthetized and Awake Rats," *Journal of Comparative Neurology* 229 (1984): 199–213.

69 *photograph of a rat's brain:* Constanze Lenschow et al., "Sexually Monomorphic Maps and Dimorphic Responses in Rat Genital Cortex," *Current Biology* 26 (2016): 106-13.

70 *rats use their whiskers:* Evgeny Bobrov et al., "The Representation of Social Facial Touch in Rat Barrel Cortex," *Current Biology* 24 (2014): 109–15.

4. OUT OF THE ETHER: BRAIN MAPS FOR HEARING

71 *Gerald Shea awoke:* Gerald Shea, *Song Without Words: Discovering My Deafness Halfway Through Life* (Boston: Da Capo Press, 2013).

76 *in a continuous map:* Melissa Saenz & Dave R. M. Langers, "Tonotopic Mapping of Human Auditory Cortex," *Hearing Research* 307 (2014): 42–52.

to hear buzzing or whistling: Wilder Penfield & Phanor Perot, "The Brain's Record of Auditory and Visual Experience: A Final Summary and Discussion," *Brain* 86 (1963): 595–696.

A1 fills in unexplained: Christopher Petkov et al., "Encoding of Illusory Continuity in Primary Auditory Cortex," *Neuron* 54 (2007): 153–65; Lars Riecke et al., "Hearing Illusory Sounds in Noise: Sensory-Perceptual Transformations in Primary Auditory Cortex," *Journal of Neuroscience* 27 (2007): 12684–89.

79 *close attention to people's lips:* Shea, *Song Without Words.*

80 *Adult rats chat with:* Heesoo Kim & Shaowen Bao, "Experience-Dependent Overrepresentation of Ultrasonic Vocalization Frequencies in the Rat Primary Auditory Cortex," *Journal of Neurophysiology* 110 (2013): 1087–96.

81 *sonic pulses and their echoes:* John E. Hill & James D. Smith, *Bats: A Natural History* (Austin: University of Texas Press, 1984).

bats have a special region: Nobuo Suga & William E. O'Neill, "Neural Axis Representing Target Range in the Auditory Cortex of the Mustache Bat," *Science* 206 (1979): 351–53.

5. BRAIN MAPS AND CODES FOR TASTE AND SMELL

87 *Your entire sense of taste:* Jayaram Chandrashekar et al., "The Receptors and Cells for Mammalian Taste," *Nature* 444 (2006): 288–94.

fooled by other molecules: David A. Yarmolinsky et al., "Common Sense About Taste: From Mammals to Insects," *Cell* 139 (2009): 234–44.

get these building blocks: Nak-Eon Choi & Jung H. Han, *How Flavor Works: The Science of Taste and Aroma* (West Sussex, UK: Wiley Blackwell, 2015).

great lengths to ingest: Choi & Han, *How Flavor Works.*

taste receptors for salt: Jayaram Chandrashekar et al., "The Cells and Peripheral Representation of Sodium Taste in Mice," *Nature* 464 (2010): 297–302.

88 *varieties of bitter-taste:* Ken L. Mueller et al., "The Receptors and Coding Logic for Bitter Taste," *Nature* 434 (2005): 225–29.

89 *grew the human version:* Mueller et al., "Receptors and Coding Logic."

versions of the sweet-taste: Grace Q. Zhao et al., "The Receptors for Mammalian Sweet and Umami Taste," *Cell* 115 (2003): 255–66.

mouse's sweet-taste receptor: Mueller et al., "Receptors and Coding Logic."

seventy-five-year-old woman: Tara M. Dutta et al., "Altered Taste and Stroke: A Case Report and Literature Review," *Topics in Stroke Rehabilitation* 20 (2013): 78–86.

91 *lies within the insula:* Dana M. Small, "Taste Representation in the Human Insula," *Brain Structure and Function* 214 (2010): 551–61.

nasty, metallic, or acidic: Laure Mazzola et al., "Gustatory and Olfactory Responses to Stimulation of the Human Insula," *Annals of Neurology* 82 (2017): 360–70.

glimpse of a taste map: Xiaoke Chen et al., "A Gustotopic Map of Taste Qualities in the Mammalian Brain," *Science* 333 (2011): 1262–66.

92 *mouse's experience of taste:* Yueqing Peng et al., "Sweet and Bitter Taste in the Brain of Awake Behaving Animals," *Nature* 527 (2015): 512–15.

93 *taste districts of the map:* Riccardo Accolla et al., "Differential Spatial Representation of Taste Modalities in the Rat Gustatory Cortex," *Journal of Neuroscience* 27 (2007): 1396–404; Max L. Fletcher et al., "Overlapping Representation of Primary Tastes in a Defined Region of the Gustatory Cortex," *Journal of Neuroscience* 37 (2017): 7595–605.

other properties of food: Takamitsu Hanamori et al., "Responses of

Neurons in the Insular Cortex to Gustatory, Visceral, and Nociceptive Stimuli in Rats," *Journal of Neurophysiology* 79 (1998): 2535–45.

these zones overlap: Mircea A. Schoenfeld et al., "Functional Magnetic Resonance Tomography Correlates of Taste Perception in the Human Primary Taste Cortex," *Neuroscience* 127 (2004): 347–53; Anna Prinster et al., "Cortical Representation of Different Taste Modalities on the Gustatory Cortex: A Pilot Study," *PLoS ONE* 12 (2017): e0190164.

with fancier techniques: Jason A. Avery et al., "Taste Quality Representation in the Human Brain," *Journal of Neuroscience* 40 (2020): 1042–52; Junichi Chikazoe et al., "Distinct Representation of Basic Taste Qualities in Human Gustatory Cortex," *Nature Communications* 10 (2019).

96 *Albatrosses and other:* Gabrielle A. Nevitt & Francesco Bonadonna, "Sensitivity to Dimethyl Sulphide Suggests a Mechanism for Olfactory Navigation by Seabirds," *Biology Letters* 1 (2005): 303–5.

ten million smell receptors: Anat Arzi & Noam Sobel, "Olfactory Perception as a Compass for Olfactory Neural Maps," *Trends in Cognitive Sciences* 15 (2011): 537–45.

98 *different patterns of activity:* Dan D. Stettler & Richard Axel, "Representations of Odor in the Piriform Cortex," *Neuron* 63 (2009): 854–64.

catfish sense of smell: Alexander A. Nikonov et al., "Beyond the Olfactory Bulb: An Odotopic Map in the Forebrain," *Proceedings of the National Academy of Sciences of the United States of America* 102 (2005): 18688–93.

100 *innate, instinctual responses:* Cory M. Root et al., "The Participation of Cortical Amygdala in Innate, Odour-Driven Behaviour," *Nature* 515 (2014): 269–73.

animal to actual odors: Root et al., "Participation of Cortical Amygdala."

101 *mice of the opposite sex:* Joseph F. Bergan et al., "Sex-Specific Processing of Social Cues in the Medial Amygdala," *eLife* 3 (2014): e02743.

odor maps in humans: Arzi & Sobel, "Olfactory Perception."

102 *humans rely on reason:* Francis Schiller, *Paul Broca: Founder of French Anthropology, Explorer of the Brain* (Berkeley: University of California Press, 1979); John P. McGann, "Poor Human Olfaction is a 19th-Century Myth," *Science* 356 (2017): eaam7263.

at the expense of smell: Grafton Elliot Smith, *The Evolution of Man: Essays* (London: Oxford University Press, 1924).

103 *humans best mice:* Amir Sarrafchi et al., "Olfactory Sensitivity for Six Predator Odorants in CD-1 Mice, Human Subjects, and Spider Monkeys," *PLoS ONE* 8 (2013): e80621; McGann, "Poor Human Olfaction."

actually count neurons: Pedro F. M. Ribeiro et al., "Greater Addition of Neurons to the Olfactory Bulb Than to the Cerebral Cortex of Eulipotyphlans but Not Rodents, Afrotherians, or Primates," *Frontiers in Neuroanatomy* 8 (2014): 23; McGann, "Poor Human Olfaction."

104 *Among the Jahai:* Asifa Majid & Niclas Burenhult, "Odors Are Expressible in Language, As Long As You Speak the Right Language," *Cognition* 130 (2014): 266–70.

bested their neighbors: Asifa Majid & Nicole Kruspe, "Hunter-Gatherer Olfaction Is Special," *Current Biology* 28 (2018): 409–13.

105 *secretions are oozing:* D. Michael Stoddart, *The Scented Ape: The Biology and Culture of Human Odour* (Cambridge, UK: Cambridge University Press, 1990).

tons of information: Mariella Pazzaglia, "Body and Odors: Not Just Molecules After All," *Current Directions in Psychological Science* 24 (2015): 329–33; Mats J. Olsson et al., "The Scent of Disease: Human Body Odor Contains an Early Chemosensory Cue of Sickness," *Psychological Science* 25 (2014): 817–23; Katrin T. Lübke & Bettina M. Pause, "Always Follow Your Nose: The Functional Significance of Social Chemosignals in Human Reproduction and Survival," *Hormones and Behavior* 68 (2015): 134–44.

your fearful sweat: Wen Zhou & Denise Chen, "Fear-Related Chemosignals Modulate Recognition of Fear in Ambiguous Facial Expressions," *Psychological Science* 20 (2009): 177–83.

bungle a procedure: Preet Bano Singh et al., "Smelling Anxiety Chemosignals Impairs Clinical Performance of Dental Students," *Chemical Senses* 43 (2018): 411–17.

106 *their menstrual cycles:* Kathleen Stern & Martha K. McClintock, "Regulation of Ovulation by Human Pheromones," *Nature* 392 (1998): 177–79.

scent of human tears: Shani Gelstein et al., "Human Tears Contain a Chemosignal," *Science* 331 (2011): 226–30.

to sniff their fingers: Idan Frumin et al., "A Social Chemosignaling Function for Human Handshaking," *eLife* 4 (2015): e05154.

6. ON THE MOVE: BRAIN MAPS FOR ACTION

107 *started in his left big toe:* John Hughlings Jackson, *Selected Writings of John Hughlings Jackson, Volume 1,* James Taylor, Gordon Holmes, & Francis Walshe, eds. (New York: Basic Books, 1958).

nine-year-old Elizabeth F.: Jackson, *Selected Writings, Volume 1.*

There was James R.: John Hughlings Jackson, "Report of a Case of Disease of One Lobe of the Cerebrum, and of Both Lobes of the Cerebellum," *Medical Mirror* (September 1, 1869): 126–27.

108 *triggered by coughing fits:* John Hughlings Jackson, "A Series of Cases Illustrative of Cerebral Pathology: Cases of Intra-Cranial Tumour," *Medical Times and Gazette* (November 30, 1872): 597–99.

keenly observant clinician: Macdonald Critchley & Eileen A. Critchley, *John Hughlings Jackson: Father of English Neurology* (New York: Oxford University Press, 1998).

"When fits begins [sic]": John Hughlings Jackson, "Case of Epileptiform Seizure, Beginning in the Right Hand," *Medical Times and Gazette* (December 23, 1871): 767–69.

111 *the frontal lobes of dogs:* Charles G. Gross, "The Discovery of Motor Cortex and Its Background," *Journal of the History of the Neurosciences* 16 (2007): 320–31.

When she died after: Jackson, "A Series of Cases Illustrative."

impact on medicine: Dee J. Canale, "William MacEwen and the Treatment of Brain Abscesses: Revisited After One Hundred Years," *Journal of Neurosurgery* 84 (1996): 133–42.

"The brain was a dark": William MacEwen, "An Address on the Surgery of the Brain and Spinal Cord," *British Medical Journal* (1888): 302–9.

Each one started with: MacEwen, "An Address on the Surgery."

112 *Penfield and his colleague:* Penfield & Boldrey, "Somatic Motor and Sensory."

113 *"intelligent and cooperative":* Penfield & Boldrey, "Somatic Motor and Sensory."

115 *their tiny electrodes:* Michael S. A. Graziano et al., "Complex Movements Evoked by Microstimulation of Precentral Cortex," *Neuron* 34 (2002): 841–51.

116 *"The stimulating electrode":* Wilder Penfield & Keasley Welch, "The Supplementary Motor Area of the Cerebral Cortex," *Archives of Neurology & Psychiatry* 66 (1951): 289–317.

117 *monkey's right motor cortex:* Graziano et al., "Complex Movements."

"The first time we found": Michael S. A. Graziano, "Ethological Action Maps: A Paradigm Shift for the Motor Cortex," *Trends in Cognitive Sciences* 20 (2016): 121–32.

118 *monkeys move naturally:* Michael S. A. Graziano et al., "Distribution of Hand Location in Monkeys During Spontaneous Behavior," *Experimental Brain Research* 155 (2004): 30–36.

120 *macaque monkey has large:* Graziano et al., "Distribution of Hand Location."

motor cortex in the mouse: Kelly A. Tennant et al., "The Organization of the Forelimb Representation of the C57BL/6 Mouse Motor Cortex as Defined by Intracortical Microstimulation and Cytoarchitecture," *Cerebral Cortex* 21 (2011): 865–76; Gustavo Arriaga et al., "Of Mice, Birds, and Men: The Mouse Ultrasonic Song System Has Some Features Similar to Humans and Song-Learning Birds," *PLoS ONE* 7 (2012): e46610.

121 *evidence of hand-to-mouth:* Michel Desmurget et al., "Neural Representations of Ethologically Relevant Hand/Mouth Synergies in the Human Precentral Gyrus," *Proceedings of the National Academy of Sciences of the United States of America* 111 (2014): 5718–22.

123 *representations for speech:* Kristofer E. Bouchard et al., "Functional Organization of Human Sensorimotor Cortex for Speech Articulation," *Nature* 495 (2013): 327–32.

125 *The elderly woman was:* Sarah Barbara Elisa Debray & Jelle Demeestere, "Alien Hand Syndrome," *Neurology* 91 (2018): 527.

127 *directions to the motor:* Martin I. Sereno & Ruey-Song Huang, "Multisensory Maps in Parietal Cortex," *Current Opinion in Neurobiology* 24 (2014): 39–46.

and another, grasping: Christina S. Konen et al., "Functional Organization of Human Posterior Parietal Cortex: Grasping- and Reaching-Related Activations Relative to Topographically Organized Cortex," *Journal of Neurophysiology* 109 (2013): 2897–908.

align visual information: Richard A. Andersen & Christopher A. Buneo, "Intentional Maps in Posterior Parietal Cortex," *Annual Review of Neuroscience* 25 (2002): 189–220.

128 *amped up or tamped down:* Aaron P. Batista et al., "Reach Plans in Eye-Centered Coordinates," *Science* 285 (1999): 257–60.

parietal cortex is injured: Jonathan R. Whitlock, "Posterior Parietal Cortex," *Current Biology* 27 (2017): R681–R701.

129 *as maps of intention:* Andersen & Buneo, "Intentional Maps"; Jacqueline Gottlieb, "From Thought to Action: The Parietal Cortex as a Bridge Between Perception, Action, and Cognition," *Neuron* 53 (2007): 9–16.

chemicals or temperature: Iwona Stepniewska et al., "Effects of Muscimol Inactivations of Functional Domains in Motor, Premotor, and Posterior Parietal Cortex on Complex Movements Evoked by Electrical Stimulation," *Journal of Neurophysiology* 111 (2014): 1100–119.

motor cortex goes rogue: Frédéric Assal et al., "Moving With or Without Will: Functional Neural Correlates of Alien Hand Syndrome," *Annals of Neurology* 62 (2007): 301–6.

7. MAPS IN THE MAKING: HOW BRAIN MAPS DEVELOP AND ADAPT

132 *proto-brain already contained:* Andrew D. Huberman et al., "Mechanisms Underlying Development of Visual Maps and Receptive Fields," *Annual Review of Neuroscience* 31 (2008): 479–509.

waves in the retina: James B. Ackman et al., "Retinal Waves Coordinate Patterned Activity Throughout the Developing Visual System," *Nature* 490 (2012): 219–25.

133 *in the fetal cochlea:* Nicolas X. Tritsch et al., "The Origin of Spontaneous Activity in the Developing Auditory System," *Nature* 450 (2007): 50–55.

touch map was refining: Roustem Khazipov & Mathieu Milh, "Early Patterns of Activity in the Developing Cortex: Focus on the Sensorimotor System," *Seminars in Cell & Developmental Biology* 76 (2018): 120–29; Shuming An et al., "Sensory-Evoked and Spontaneous Gamma and Spindle Bursts in Neonatal Rat Motor Cortex," *Journal of Neuroscience* 34 (2014): 10870–83.

134 *highways of connections:* Julien Dubois et al., "The Early Development of Brain White Matter: A Review of Imaging Studies in Fetuses, Newborns, and Infants," *Neuroscience* 276 (2014): 48–71.

135 *When the pups are born:* Heesoo Kim & Shaowen Bao, "Experience-Dependent Overrepresentation of Ultrasonic Vocalization Frequencies in the Rat Primary Auditory Cortex," *Journal of Neurophysiology* 110 (2013): 1087–96.

around 50,000 hertz: Markus Wöhr, "Ultrasonic Communication in Rats: Appetitive 50-kHz Ultrasonic Vocalizations as Social Contact Calls," *Behavioral Ecology and Sociobiology* 72 (2018): 14.

the rat pup's A1 map: Kim & Bao, "Experience-Dependent Overrepresentation."

137 *placed in a sound chamber:* Li I. Zhang et al., "Persistent and Specific Influences of Early Acoustic Environments on Primary Auditory Cortex," *Nature Neuroscience* 4 (2001): 1123–30.

139 *sound and touch processing:* Cristina Baldoli et al., "Maturation of Preterm Newborn Brains: An fMRI-DTI Study of Auditory Processing of Linguistic Stimuli and White Matter Development," *Brain Structure*

and Function 220 (2015): 3733–51; Rebeccah Slater et al., "Premature Infants Display Increased Noxious-Evoked Neuronal Activity in the Brain Compared to Healthy Age-Matched Term-Born Infants," *NeuroImage* 52 (2010): 583–89; Johanna Hohmeister et al., "Cerebral Processing of Pain in School-Aged Children with Neonatal Nociceptive Input: An Exploratory fMRI Study," *Pain* 150 (2010): 257–67.

140 *the mother's heartbeat:* Alexandra R. Webb et al., "Mother's Voice and Heartbeat Sounds Elicit Auditory Plasticity in the Human Brain Before Full Gestation," *Proceedings of the National Academy of Sciences of the United States of America* 112 (2015): 3152–57.

first few weeks after birth: Eileen E. Birch et al., "The Critical Period for Surgical Treatment of Dense Congenital Bilateral Cataracts," *Journal of the American Association for Pediatric Ophthalmology & Adult Strabismus* 13 (2009): 67–71.

had cataracts as infants: Daphne Maurer, "Critical Periods Re-examined: Evidence from Children Treated for Dense Cataracts," *Cognitive Development* 42 (2017): 27–36.

142 *redistrict their visual cortex:* Amir Amedi et al., "The Occipital Cortex in the Blind: Lessons About Plasticity and Vision," *Current Directions in Psychological Science* 14 (2005): 306–11; Marina Bedny, "Evidence from Blindness for a Cognitively Pluripotent Cortex," *Trends in Cognitive Sciences* 21 (2017): 637–48.

more dramatically remapped: Shipra Kanjlia et al., "Sensitive Period for Cognitive Repurposing of Human Visual Cortex," *Cerebral Cortex* (2018). doi: 10.1093/cercor/bhy280.

143 *adult rats were exposed:* Richard G. Rutkowski & Norman M. Weinberger, "Encoding of Learned Importance of Sound by Magnitude of Representation Area in Primary Auditory Cortex," *Proceedings of the National Academy of Sciences of the United States of America* 102 (2005): 13664–69.

145 *community in western Kenya:* Charles M. Super, "Environmental Effects on Motor Development: The Case of 'African Infant Precocity,'" *Developmental Medicine and Child Neurology* 18 (1976): 561–67.

146 *brain scans to measure:* Katrin Amunts et al., "Motor Cortex and Hand Motor Skills: Structural Compliance in the Human Brain," *Human Brain Mapping* 5 (1997): 206–15.

147 *for only the left hand:* Thomas Elbert et al., "Increased Cortical Representation of the Fingers of the Left Hand in String Players," *Science* 270 (1995) 305–7.

individual weekly piano lessons: Krista L. Hyde et al., "Musical Training Shapes Structural Brain Development," *Journal of Neuroscience* 29 (2009): 3019–25.

8. KNOWING AGAIN: BRAIN MAPS FOR RECOGNITION

152 *"I have not even the"*: Glyn W. Humphreys & Jane M. Riddoch, *To See but Not to See: A Case Study of Visual Agnosia* (London: Lawrence Erlbaum Associates, 1987).

"A carrot is a root": Humphreys & Riddoch, *To See but Not to See.*

153 *"He cannot find his way"*: Humphreys & Riddoch, *To See but Not to See.*

"I cannot recognize": Humphreys & Riddoch, *To See but Not to See.*

154 *have noticed a pattern:* Elizabeth Warrington & Tim Shallice, "Category Specific Semantic Impairments," *Brain* 107 (1984): 829–54; Iftah Biran & H. Branch Coslett, "Visual Agnosia," *Current Neurology and Neuroscience Reports* 3 (2003): 508–12.

155 *face recognition possible:* Nancy Kanwisher et al., "The Fusiform Face Area: A Module in Human Extrastriate Cortex Specialized for Face Perception," *Journal of Neuroscience* 17 (1997): 4302–11.

"An initial scan with me": Nancy Kanwisher, "The Quest for the FFA and Where It Led," *Journal of Neuroscience* 37 (2017): 1056–61.

156 *An estimated 2 percent:* Ingo Kennerknecht et al., "Prevalence of Hereditary Prosopagnosia (HPA) in Hong Kong Chinese Population," *American Journal of Medical Genetics Part A* 146A (2008): 2863–70.

subtle structural differences: Sunbin Song et al., "Local but Not Long-Range Microstructural Differences of the Ventral Temporal Cortex in Developmental Prosopagnosia," *Neuropsychologia* 78 (2015): 195–206.

157 *to undergo brain surgery:* Josef Parvizi et al., "Electrical Stimulation of Human Fusiform Face-Selective Regions Distorts Face Perception," *Journal of Neuroscience* 32 (2012): 14915–20; quotations transcribed from supplementary movies published with the article.

159 *area in the occipital cortex:* David Pitcher et al., "The Role of the Occipital Face Area in the Cortical Face Perception Network," *Experimental Brain Research* 209 (2011): 481–93.

a particular environment: Russell Epstein & Nancy Kanwisher, "A Cortical Representation of the Local Visual Environment," *Nature* 392 (1998): 598–601.

unable to recognize landmarks: Geoffrey K. Aguirre & Mark D'Esposito, "Topographical Disorientation: A Synthesis and Taxonomy," *Brain* 122 (1999) 1613–28.

A medical team did this: Pierre Mégevand et al., "Seeing Scenes: Topographic Visual Hallucinations Evoked by Direct Electrical Stimulation of the Parahippocampal Place Area," *Journal of Neuroscience* 34 (2014): 5399–405.

160 *pictures of bodies:* Paul E. Downing et al., "A Cortical Area Selective for Visual Processing of the Human Body," *Science* 293 (2001): 2470–73.

about a body's shape: Paul E. Downing & Marius V. Peelen, "Body Selectivity in Occipitotemporal Cortex: Causal Evidence," *Neuropsychologia* 83 (2016): 138–48.

the fusiform body area: Rebecca F. Schwarzlose et al., "Separate Face and Body Selectivity on the Fusiform Gyrus," *Journal of Neuroscience* 25 (2005): 11055–59.

of different body parts: Valentina Moro et al., "The Neural Basis of Body Form and Body Action Agnosia," *Neuron* 60 (2008): 235–46.

161 *specialize in handheld tools:* James W. Lewis, "Cortical Networks Related to Human Use of Tools," *Neuroscientist* 12 (2006): 211–31.

shapes of written letters: Laurent Cohen et al., "Language-Specific Tuning of Visual Cortex? Functional Properties of the Visual World Form Area," *Brain* 125 (2002): 1054–69.

163 *overarching map is organized:* Talia Konkle & Alfonso Caramazza, "Tripartite Organization of the Ventral Stream by Animacy and Object Size," *Journal of Neuroscience* 33 (2013): 10235–42.

164 *offer us specific opportunities:* James J. Gibson, *The Ecological Approach to Visual Perception* (Boston: Houghton Mifflin, 1979).

165 *effect of these weak waves:* Michael J. Arcaro & Margaret S. Livingstone, "A Hierarchical, Retinotopic Proto-organization of the Primate Visual System at Birth," *eLife* 6 (2017): e26196.

younger than three months: Linda B. Smith et al., "The Developing Infant Creates a Curriculum for Statistical Learning," *Trends in Cognitive Sciences* 22 (2018): 325–36.

166 *macaques have an object map:* Doris Y. Tsao et al., "Faces and Objects in Macaque Cerebral Cortex," *Nature Neuroscience* 6 (2003): 989–95.

pictures to newborn macaques: Margaret S. Livingstone et al., "Development of the Macaque Face-Patch System," *Nature Communications* 8 (2017): 10.1038/ncomms14897.

three monkeys from birth: Michael J. Arcaro et al., "Seeing Faces Is Necessary for Face-Domain Formation," *Nature Neuroscience* 20 (2017): 1404–12.

167 *babies four to six months old:* Ben Deen et al., "Organization of High-

Level Visual Cortex in Human Infants," *Nature Communications* 8 (2017): 13995.

privileged categories: Kalanit Grill-Spector et al., "Developmental Neuroimaging of the Ventral Visual Cortex," *Trends in Cognitive Sciences* 12 (2008): 152–62; Golijeh Golarai et al., "Experience Shapes the Development of Neural Substrates of Face Processing in Human Ventral Temporal Cortex," *Cerebral Cortex* 27 (2015): bhv314.

babies whose left eye: Richard Le Grand et al., "Expert Face Processing Requires Visual Input to the Right Hemisphere During Infancy," *Nature Neuroscience* 6 (2003): 1108–12.

168 *far more views of hands:* Caitlin M. Fausey et al., "From Faces to Hands: Changing Visual Input in the First Two Years," *Cognition* 152 (2016): 101–7.

develop a new zone: Stanislas Dehaene et al., "Illiterate to Literate: Behavioural and Cerebral Changes Induced by Reading Acquisition," *Nature Reviews Neuroscience* 16 (2015): 234–44.

169 *scientists trained adults:* Jill Weisberg et al., "A Neural System for Learning about Object Function," *Cerebral Cortex* 17 (2007): 513–21.

are by no means limited: Marius V. Peelen & Paul E. Downing, "Category Selectivity in Human Visual Cortex: Beyond Visual Object Recognition," *Neuropsychologia* 105 (2017): 177–83.

170 *similar to those of sighted:* Job van den Hurk et al., "Development of Visual Category Selectivity in Ventral Visual Cortex Does Not Require Visual Experience," *Proceedings of the National Academy of Sciences of the United States of America* 114 (2017): E4501–E4510; Xiaoying Wang et al., "How Visual Is the Visual Cortex? Comparing Connectional and Functional Fingerprints Between Congenitally Blind and Sighted Individuals," *Journal of Neuroscience* 35 (2015): 12545–59.

9. IMAGINING, REMEMBERING, AND PAYING ATTENTION WITH BRAIN MAPS

171 *Penfield referred to her:* Wilder Penfield, *The Excitable Cortex in Conscious Man* (Springfield, IL: C. C. Thomas, 1958).

172 *"feeling—as though I":* Penfield, *The Excitable Cortex.*

"I think I heard a mother": Penfield, *The Excitable Cortex.*

173 *about half our waking hours:* Matthew A. Killingsworth & Daniel T. Gilbert, "A Wandering Mind Is an Unhappy Mind," *Science* 330 (2010): 932.

174 *"Think of some definite"*: Francis Galton, *Inquiries into Human Faculty and Its Development* (London: Macmillan, 1883).

175 *"Call up before your"*: Galton, *Inquiries into Human Faculty.*

176 *psychology and neuroscience:* Stephen Jay Gould, *The Mismeasure of Man,* 2nd ed. (New York: W. W. Norton, 1996).

177 *a person's mental imagery:* Joel Pearson & Stephen M. Kosslyn, "The Heterogeneity of Mental Representation: Ending the Imagery Debate," *Proceedings of the National Academy of Sciences of the United States of America* 33 (2015): 10089–92.

visual map nicknamed V2: Anke Marit Albers et al., "Shared Representations for Working Memory and Mental Imagery in Early Visual Cortex," *Current Biology* 23 (2013): 1427–31.

Imagining visual motion: Amanda Kaas et al., "Imagery of a Moving Object: The Role of Occipital Cortex and Human MT/V5+," *NeuroImage* 49 (2010): 794–804.

Imagining a face: Kathleen M. O'Craven & Nancy Kanwisher, "Mental Imagery of Faces and Places Activates Corresponding Stimulus-Specific Brain Regions," *Journal of Cognitive Neuroscience* 12 (2000): 1013–23.

weaker form of actually seeing: Joel Pearson et al., "Mental Imagery: Functional Mechanisms and Clinical Applications," *Trends in Cognitive Sciences* 19 (2015): 590–602.

178 *When you imagine a sound:* Robert J. Zatorre & Andrea R. Halpern, "Mental Concerts: Musical Imagery and Auditory Cortex," *Neuron* 47 (2005): 9–12.

imagine being touched: Nan J. Wise et al., "Activation of Sensory Cortex by Imagined Genital Stimulation: An fMRI Analysis," *Socioaffective Neuroscience & Psychology* 6 (2016): 31481.

When you imagine speaking: Kayoko Okada et al., "Neural Evidence for Predictive Coding in Auditory Cortex During Speech Production," *Psychonomic Bulletin & Review* 25 (2018): 423–30.

imagine moving your fingers: Carlo A. Porro et al., "Primary Motor and Sensory Cortex Activation During Motor Performance and Motor Imagery: A Functional Magnetic Resonance Imaging Study," *Journal of Neuroscience* 16 (1996): 7688–98.

179 *activity in brain maps:* Mariam R. Sood & Martin I. Sereno, "Areas Activated During Naturalistic Reading Comprehension Overlap Topological Visual, Auditory, and Somatomotor Maps," *Human Brain Mapping* 37 (2016): 2784–810.

this recollected imagery: Denis Le Bihan et al., "Activation of Human Primary Visual Cortex During Visual Recall: A Magnetic Resonance

Imaging Study," *Proceedings of the National Academy of Sciences of the United States of America* 90 (1993): 11802–5.

intrusive and upsetting imagery: Emily A. Holmes et al., "Mental Imagery in Depression: Phenomenology, Potential Mechanisms, and Treatment Implications," *Annual Review of Clinical Psychology* 12 (2016): 249–80; Pearson et al., "Mental Imagery."

180 *several intriguing observations:* Alan Baddeley, "Working Memory," *Science* 255 (1992): 556–59.

speaking and hearing the content: Bradley R. Buchsbaum & Mark D'Esposito, "The Search for the Phonological Store: From Loop to Convolution," *Journal of Cognitive Neuroscience* 20 (2008): 762–78; Stefan Koelsch et al., "Functional Architecture of Verbal and Tonal Working Memory: An fMRI Study," *Human Brain Mapping* 30 (2009): 859–73.

181 *form of working memory:* Baddeley, "Working Memory."

Looking at an image: Albers et al., "Shared Representations."

dream of seeing a face: Francesca Siclari et al., "The Neural Correlates of Dreaming," *Nature Neuroscience* 20 (2017): 872–78.

182 *have smaller receptive fields:* Ben M. Harvey & Serge O. Dumoulin, "The Relationship Between Cortical Magnification Factor and Population Receptive Field Size in Human Visual Cortex: Constancies in Cortical Architecture," *Journal of Neuroscience* 31 (2011): 13604–12.

people with larger V1 maps: Johanna Bergmann et al., "Neural Anatomy of Primary Visual Cortex Limits Visual Working Memory," *Cerebral Cortex* 26 (2016): 43–50; Johanna Bergmann et al., "Smaller Primary Visual Cortex Is Associated with Stronger, but Less Precise Mental Imagery," *Cerebral Cortex* 26 (2016): 3838–50.

"To my astonishment": Galton, *Inquiries into Human Faculty.*

visual imagery is affected: Martha J. Farah, "Is Visual Imagery Really Visual? Overlooked Evidence from Neuropsychology," *Psychological Review* 95 (1988): 307–17.

183 *disrupting neural activity:* Stephen M. Kosslyn et al., "The Role of Area 17 in Visual Imagery: Convergent Evidence from PET and rTMS," *Science* 284 (1999): 167–70.

This sixty-five-year-old man: Adam Z. J. Zeman et al., "Loss of Imagery Phenomenology with Intact Visuo-Spatial Task Performance: A Case of 'Blind Imagination,'" *Neuropsychologia* 48 (2010): 145–55.

to generate mental images: Adam Zeman et al., "Lives Without Imagery—Congenital Aphantasia," *Cortex* 73 (2015): 378–80.

their teens or twenties: Zeman et al., "Lives Without Imagery."

184 *brain can be co-opted:* Michael L. Anderson, "Neural Reuse: A Fun-

damental Organizational Principle of the Brain," *Behavioral and Brain Sciences* 33 (2010): 245–313.

imagery and perception interfere: Cheves W. Perky, "An Experimental Study of Imagination," *American Journal of Psychology* 21 (1910): 422–52; Alumit Ishai & Dov Sagi, "Visual Imagery Facilitates Visual Perception: Psychophysical Evidence," *Journal of Cognitive Neuroscience* 9 (1997): 476–89.

186 *"Everyone knows what attention":* William James, *The Principles of Psychology* (New York: H. Holt & Company, 1890).

188 *items at your center of gaze:* David C. Somers et al., "Functional MRI Reveals Spatially Specific Attentional Modulation in Human Primary Visual Cortex," *Proceedings of the National Academy of Sciences of the United States of America* 96 (1999): 1663–68.

Attending to faces: Ewa Wojciulik et al., "Covert Visual Attention Modulates Face-Specific Activity in the Human Fusiform Gyrus: fMRI Study," *Journal of Neurophysiology* 79 (1998): 1574–78.

189 *whether a fluid tastes sweet:* Maria G. Veldhuizen et al., "Trying to Detect Taste in a Tasteless Solution: Modulation of Early Gustatory Cortex by Attention to Taste," *Chemical Senses* 32 (2007): 569–81.

substantially more complex: Tirin Moore & Marc Zirnsak, "Neural Mechanisms of Selective Visual Attention," *Annual Review of Psychology* 68 (2017): 47–72; John H. Reynolds et al., "Attention Increases Sensitivity of V4 Neurons," *Neuron* 26 (2000): 703–14; Thomas C. Sprague et al., "Visual Attention Mitigates Information Loss in Small- and Large-Scale Neural Codes," *Trends in Cognitive Sciences* 19 (2015): 215–26.

regions of the sensory maps: Thomas C. Sprague & John T. Serences, "Attention Modulates Spatial Priority Maps in the Human Occipital, Parietal, and Frontal Cortices," *Nature Neuroscience* 16 (2013): 1879–87; Moore & Zirnsak, "Neural Mechanisms."

10. COMPREHENDING AND COMMUNICATING WITH BRAIN MAPS

192 *approximate number awareness:* Andreas Nieder & Stanislas Dehaene, "Representation of Number in the Brain," *Annual Review of Neuroscience* 32 (2009): 185–208.

young as three months old: Véronique Izard et al., "Distinct Cerebral Pathways for Object Identity and Number in Human Infants," *PLoS Biology* 6 (2008): e11.

maps of approximate number: Ben M. Harvey & Serge O. Dumoulin, "A Network of Topographic Numerosity Maps in Human Association Cortex," *Nature Human Behaviour* 1 (2017): 0036.

193 *between numbers and space:* Edward M. Hubbard et al., "Interactions Between Number and Space in Parietal Cortex," *Nature Reviews Neuroscience* 6 (2005): 435–48.

birds, monkeys, and human infants: Elizabeth Y. Toomarian & Edward M. Hubbard, "On the Genesis of Spatial-Numerical Associations: Evolutionary and Cultural Factors Co-construct the Mental Number Line," *Neuroscience & Biobehavioral Reviews* 90 (2018): 184–99.

associate smaller numbers: Stanislas Dehaene et al., "The Mental Representation of Parity and Number Magnitude," *Journal of Experimental Psychology: General* 122 (1993): 371–96.

the task and the context: Daniel Bächtold et al., "Stimulus-Response Compatibility in Representational Space," *Neuropsychologia* 36 (1998): 731–35; Mattias Hartmann et al., "There Is More Than 'More Is Up': Hand and Foot Responses Reverse the Vertical Association of Number Magnitudes," *Journal of Experimental Psychology: Human Perception and Performance* 40 (2014): 1401–14.

194 *seeing small numbers:* Hubbard et al., "Interactions Between."

195 *a simple number task:* Marco Zorzi et al., "Neglect Disrupts the Mental Number Line," *Nature* 417 (2002): 138–39; Patrik Vuilleumier et al., "The Number Space and Neglect," *Cortex* 40 (2004): 399–410.

"marked inability to handle": Elena Rusconi & Roberto Cubelli, "The Making of a Syndrome: The English Translation of Gerstmann's First Report," *Cortex* 117 (2019): 277–83.

196 *"The fingers have lost their":* Marcel Kinsbourne & Elizabeth K. Warrington, "A Study of Finger Agnosia," *Brain* 85 (1962): 47–66.

temporarily hush the activity: Elena Rusconi et al., "Dexterity with Numbers: rTMS over Left Angular Gyrus Disrupts Finger Gnosis and Number Processing," *Neuropsychologia* 43 (2005): 1609–24.

stimulated this parietal region: Franck-Emmanuel Roux et al., "Writing, Calculating, and Finger Recognition in the Region of the Angular Gyrus: A Cortical Stimulation Study of Gerstmann Syndrome," *Journal of Neurosurgery* 99 (2003): 716–27.

197 *rather than identical terrain:* Elena Rusconi et al., "A Disconnection Account of Gerstmann Syndrome: Functional Neuroanatomy Evidence," *Annals of Neurology* 66 (2009): 654–62.

on finger-awareness tasks: Michel Fayol et al., "Predicting Arithmetical Achievement from Neuro-Psychological Performance: A Longitudinal Study," *Cognition* 68 (1998): B63–B70; Marie-Pascale Noël, "Finger Gnosia: A Predictor of Numerical Abilities in Children?" *Child Neuropsychology* 11 (2005): 413–30.

blind children do not tend: Virginie Crollen et al., "The Role of Vision in the Development of Finger-Number Interactions: Finger-Counting and Finger-Montring in Blind Children," *Journal of Experimental Child Psychology* 109 (2011): 525–39.

born with cerebral palsy: Nolwenn Guedin et al., "Dexterity and Finger Sense: A Possible Dissociation in Children with Cerebral Palsy," *Perceptual and Motor Skills* 125 (2018): 718–31.

198 *child of the Yupno people:* Andrea Bender & Sieghard Beller, "Cultural Variation in Numeration Systems and Their Mapping onto the Mental Number Line," *Journal of Cross-Cultural Psychology* 42 (2011): 579–97.

our concept of elapsed time: Jennifer T. Coull & Sylvie Droit-Volet, "Explicit Understanding of Duration Develops Implicitly Through Action," *Trends in Cognitive Sciences* 22 (2018): 923–37.

199 *between time and space:* Rafael Núñez & Kensy Cooperrider, "The Tangle of Space and Time in Human Cognition," *Trends in Cognitive Sciences* 17 (2013): 220–29; Lera Boroditsky, "Language and the Construction of Time Through Space," *Trends in Cognitive Sciences* 41 (2018): 651–53.

The Yupno people: Rafael Núnez et al., "Contours of Time: Topographic Construals of Past, Present, and Future in the Yupno Valley of Papua New Guinea," *Cognition* 124 (2012): 25–35.

200 *their mental time line:* Arnaud Saj et al., "Patients with Left Spatial Neglect Also Neglect the 'Left Side' of Time," *Psychological Science* 25 (2014): 207–14.

201 *how much time has elapsed:* Hugo Merchant et al., "Neural Basis of the Perception and Estimation of Time," *Annual Review of Neuroscience* 36 (2013): 313–36.

regions like the hippocampus: Howard Eichenbaum, "Time Cells in the Hippocampus: A New Dimension for Mapping Memories," *Nature Reviews Neuroscience* 15 (2014): 732–44.

202 *reading the word* kick: Olaf Hauk et al., "Somatotopic Representation of Action Words in Human Motor and Premotor Cortex," *Neuron* 41 (2004): 301–7.

patterns of activation: Ilan Dinstein et al., "Brain Areas Selective for Both Observed and Executed Movements," *Journal of Neurophysiology* 98 (2007): 1415–27.

Watching someone else: Gina Caetano et al., "Actor's and Observer's

Primary Motor Cortices Stabilize Similarly After Seen or Heard Motor Actions," *Proceedings of the National Academy of Sciences of the United States of America* 104 (2007): 9058–62.

Reading a word that: Alfonso Barrós-Loscertales et al., "Reading Salt Activates Gustatory Brain Regions: fMRI Evidence for Semantic Grounding in a Novel Sensory Modality," *Cerebral Cortex* 22 (2012): 2554–63; Julio González et al., "Reading *cinnamon* Activates Olfactory Brain Regions," *NeuroImage* 32 (2006): 906–12; Markus Kiefer et al., "The Sound of Concepts: Four Markers for a Link Between Auditory and Conceptual Brain Systems," *Journal of Neuroscience* 28 (2008): 12224–30.

204 *tip of the temporal lobe:* Matthew A. Lambon Ralph et al., "The Neural and Computational Bases of Semantic Cognition," *Nature Reviews Neuroscience* 18 (2017): 42–55.

known as semantic dementia: John R. Hodges & Karalyn Patterson, "Semantic Dementia: A Unique Clinicopathological Syndrome," *Lancet Neurology* 6 (2007): 1004–14.

205 *in other viewers' brains:* Uri Hasson et al., "Intersubject Synchronization of Cortical Activity During Natural Vision," *Science* 303 (2004): 1634–40.

these ideas a step further: Alexander G. Huth et al., "A Continuous Semantic Space Describes the Representation of Thousands of Object and Action Categories Across the Human Brain," *Neuron* 76 (2012): 1210–24.

206 *spontaneously tell a personal story:* Greg J. Stephens et al., "Speaker-Listener Neural Coupling Underlies Successful Communication," *Proceedings of the National Academy of Sciences of the United States of America* 107 (2010): 14425–30.

207 *teacher delivers a lecture:* Mai Nguyen et al., "Teacher-Student Neural Coupling During Teaching and Learning," *bioRxiv* (2020).

"Dear Mother and Daddy": John Dermot Turing, "The Man with the Terrible Trousers," in *The Turing Guide,* B. Jack Copeland et al., eds. (Oxford, UK: Oxford University Press, 2017), 21.

208 *Turing was recruited:* Andrew Hodges, *Alan Turing: The Enigma* (New York: Simon & Schuster, 1983).

11. MAPS AS PORTALS: MIND READING AND MIND WRITING WITH BRAIN MAPS

215 *Carol had not spoken:* Adrian Owen, *Into the Gray Zone: A Neuroscientist Explores the Border Between Life and Death* (New York: Scribner, 2017).

there is no single line: Tim Bayne et al., "Are There Levels of Consciousness?" *Trends in Cognitive Sciences* 20 (2016): 405–13.

216 *brain was alert and aware:* Adrian M. Owen et al., "Detecting Awareness in the Vegetative State," *Science* 313 (2006): 1402.

one in five patients: Martin M. Monti et al., "Willful Modulation of Brain Activity in Disorders of Consciousness," *New England Journal of Medicine* 362 (2010): 579–89.

217 *less expensive alternatives:* Damian Cruse et al., "Bedside Detection of Awareness in the Vegetative State: A Cohort Study," *Lancet* 378 (2011): 2088–94.

because of Juan: Owen, *Into the Gray Zone.*

218 *shown superimposed pictures:* Ewa Wojciulik et al., "Covert Visual Attention Modulates Face-Specific Activity in the Human Fusiform Gyrus: fMRI Study," *Journal of Neurophysiology* 79 (1998): 1574–78.

219 *allocation of attention:* Megan T. deBettencourt et al., "Closed-Loop Training of Attention with Real-Time Brain Imaging," *Nature Neuroscience* 18 (2015): 470–75.

220 *same small area of V1:* Haidong D. Lu & Anna W. Roe, "Functional Organization of Color Domains in V1 and V2 of Macaque Monkey Revealed by Optical Imaging," *Cerebral Cortex* 18 (2008): 516–33.

221 *(pressure, flutter, and vibration):* Robert M. Friedman et al., "Modality Maps Within Primate Somatosensory Cortex," *Proceedings of the National Academy of Sciences of the United States of America* 101 (2004): 12724–29.

radially like a pinwheel: Mark L. Andermann & Christopher I. Moore, "A Somatotopic Map of Vibrissa Motion Direction Within a Barrel Column," *Nature Neuroscience* 9 (2006): 543–51.

225 *tell with reasonable accuracy:* Albers et al., "Shared Representations"; Yukiyasu Kamitani & Frank Tong, "Decoding the Visual and Subjective Contents of the Human Brain," *Nature Neuroscience* 8 (2005): 679–85; Kendrick N. Kay et al., "Identifying Natural Images from Human Brain Activity," *Nature* 452 (2008): 352–55; Thomas Naselaris et al., "A Voxel-Wise Encoding Model for Early Visual Areas Decodes Mental Images of Remembered Scenes," *NeuroImage* 105 (2015): 215–28; Sean M. Polyn et al., "Category-Specific Cortical Activity Precedes Retrieval During Memory Search," *Science* 310 (2005): 1963–66.

226 *a person was dreaming:* Tomoyasu Horikawa et al., "Neural Decoding of Visual Imagery During Sleep," *Science* 340 (2013): 639–42.

clips of spoken sentences: Elia Formisano et al., "'Who' Is Saying

'What'? Brain-Based Decoding of Human Voice and Speech," *Science* 322 (2008): 970–73.

have proved able to decode: Kay H. Brodersen et al., "Decoding the Perception of Pain from fMRI Using Multivariate Pattern Analysis," *NeuroImage* 63 (2012): 1162–70; Tom M. Mitchell et al., "Predicting Human Brain Activity Associated with the Meanings of Nouns," *Science* 320 (2008): 1191–95; Timothy J. Vickery et al., "Ubiquity and Specificity of Reinforcement Signals Throughout the Brain," *Neuron* 72 (2011): 166–77.

studies tackled this challenge: Yoichi Miyawaki et al., "Visual Image Reconstruction from Human Brain Activity Using a Combination of Multiscale Local Image Decoders," *Neuron* 60 (2008): 915–29; Thomas Naselaris et al., "Bayesian Reconstruction of Natural Images from Human Brain Activity," *Neuron* 63 (2009): 902–15.

227 *deception and truth-telling:* Daniel D. Langleben & Jane Campbell Moriarty, "Using Brain Imaging for Lie Detection: Where Science, Law, and Research Policy Collide," *Psychology, Public Policy, and Law* 19 (2013): 222–34; Martha J. Farah et al., "Functional MRI-Based Lie Detection: Scientific and Societal Challenges," *Nature Reviews Neuroscience* 15 (2014): 123–31; Russell A. Poldrack, *The New Mind Readers: What Neuroimaging Can and Cannot Reveal About Our Thoughts* (Princeton, NJ: Princeton University Press, 2018).

imperceptible movements: Giorgio Ganis et al., "Lying in the Scanner: Covert Countermeasures Disrupt Deception Detection by Functional Magnetic Resonance Imaging," *NeuroImage* 55 (2011): 312–19.

228 *the specific human faces:* Le Chang & Doris Y. Tsao, "The Code for Facial Identity in the Primate Brain," *Cell* 169 (2017): 1013–28.

230 *control an iPad with his mind:* Paul Nuyujukian et al., "Cortical Control of a Tablet Computer by People with Paralysis," *PLoS ONE* 13 (2018): e0204566.

"the visualization I found": VICE on HBO, "The Future of Brain Hacking." Bonus posted to YouTube on September 21, 2018: https://www.youtube.com/watch?v=rfWWBB7csTo.

had electrodes implanted: A. Bolu Ajiboye et al., "Restoration of Reaching and Grasping Movements Through Brain-Controlled Muscle Stimulation in a Person with Tetraplegia: A Proof-of-Concept Demonstration," *Lancet* 389 (2017): 1821–30.

231 *"It was amazing":* Case Western Reserve University channel, "Man with Quadriplegia Employs Injury Bridging Technologies to Move Again

—Just by Thinking," posted to YouTube on March 28, 2017: https://www.youtube.com/watch?v=OHsFkqSM7-A.

232 *radio receivers and electrodes:* Brindley & Lewin, "The Sensations Produced."

233 *"I was not able to make out":* Jens Naumann, *Search for Paradise: A Patient's Account of the Artificial Vision Experiment* (Bloomington: XLibris Corporation, 2012).

"a few dots of light": Naumann, *Search for Paradise.*

234 *experimental cortical prosthetics:* Philip M. Lewis & Jeffrey V. Rosenfeld, "Electrical Stimulation of the Brain and the Development of Cortical Visual Prostheses: An Historical Perspective," *Brain Research* 1630 (2016): 208–24.

do not predictably coalesce: William H. Bosking et al., "Electrical Stimulation of Visual Cortex: Relevance for the Development of Visual Cortical Prosthetics," *Annual Review of Vision Science* 3 (2017): 141–66.

235 *dynamically tracing letters:* Michael S. Beauchamp et al., "Dynamic Stimulation of Visual Cortex Produces Form Vision in Sighted and Blind Humans," *Cell* 181 (2020): 774–83.

devices that can be carried: Peter B. Shull & Dana D. Damian, "Haptic Wearables as Sensory Replacement, Sensory Augmentation, and Trainer—A Review," *Journal of NeuroEngineering and Rehabilitation* 12 (2015). DOI 10.1186/s12984-015-0055-z.

implanted in the whisker region: Eric E. Thomson et al., "Perceiving Invisible Light Through a Somatosensory Cortical Prosthesis," *Nature Communications* 4 (2013): 1482.

237 *the functional connectivity among:* Antonello Baldassarre et al., "Individual Variability in Functional Connectivity Predicts Performance on a Perceptual Task," *Proceedings of the National Academy of Sciences of the United States of America* 109 (2012): 3516–21.

person will learn or perform: John D. E. Gabrieli et al., "Prediction as a Humanitarian and Pragmatic Contribution from Human Cognitive Neuroscience," *Neuron* 85 (2015): 11–26; "A Neuromarker of Sustained Attention from Whole-Brain Functional Connectivity," *Nature Neuroscience* 19 (2016): 165–71.

238 *type of mental illness:* Barnaly Rashid & Vince Calhoun, "Towards a Brain-Based Predictome of Mental Illness," *Human Brain Mapping* 41 (2020): 3468–535.

has been used to predict: Gabrieli et al., "Prediction as a Humanitarian."

signals recorded from babies: Dennis L. Molfese, "Predicting Dyslexia at 8 Years of Age Using Neonatal Brain Responses," *Brain and Language* 72 (2000): 238–45; Hyunseok Kook et al., "Multi-stimuli Multi-Channel Data and Decision Fusion Strategies for Dyslexia Prediction Using Neonatal ERPs," *Pattern Recognition* 38 (2005): 2174–84.

would be rearrested: Eyal Aharoni et al., "Neuroprediction of Future Rearrest," *Proceedings of the National Academy of Sciences of the United States of America* 110 (2013): 6223–28.

239 *"achieve a sort of symbiosis":* Elon Musk, Neuralink Launch Event, July 16, 2019, San Francisco: https://www.youtube.com/watch?v=r-vbh3t7WVI.

Google's parent company: GSK Press Release: "GSK and Verily to Establish Galvani Bioelectronics—A New Company Dedicated to the Development of Bioelectric Medicines," August 1, 2016: https://www.gsk.com/en-gb/media/press-releases/gsk-and-verily-to-establish-galvani-bioelectronics-a-new-company-dedicated-to-the-development-of-bioelectronic-medicines/.

quickly and accurately: David A. Moses et al., "Real-time Decoding of Question-and-Answer Speech Dialogue Using Human Cortical Activity," *Nature Communications* 10 (2019): 3096.

"We're working on a system": Mark Zuckerberg, April 19, 2017: https://www.facebook.com/zuck/videos/vb.4/10103661167577621/?type=2&theater.

240 *a teenager was pregnant:* Charles Duhigg, "How Companies Learn Your Secrets," *New York Times Magazine,* February 16, 2012.

identify the emotional states: Adam D. I. Kramer et al., "Experimental Evidence of Massive-Scale Emotional Contagion Through Social Networks," *Proceedings of the National Academy of Sciences of the United States of America* 111 (2014): 8788–90; Nitasha Tiku, "Get Ready for the Next Big Privacy Backlash Against Facebook," *Wired,* May 21, 2017.

tendency for depression: Rafail-Evangelos Mastoras et al., "Touchscreen Typing Pattern Analysis for Remote Detection of the Depressive Tendency," *Scientific Reports* 9 (2019): 13414.

work along the same lines: Luca Giancardo et al., "Computer Keyboard Interaction as an Indicator of Early Parkinson's Disease," *Scientific Reports* 6 (2016): 34468; Alicia Nieto-Reyes et al., "Classification of Alzheimer's Patients Through Ubiquitous Computing," *Sensors* 17 (2017): 1679.

241 *"The unavoidable point":* The Royal Society, *iHuman: Blurring Lines Between Mind and Machine* (2019). http://www.royalsociety.org/ihuman-perspective.

242 *concerned neuroscientists:* Rafael Yuste et al., "Four Ethical Priorities for Neurotechnologies and AI," *Nature* 551 (2017): 159–63.

12. HOW BRAIN MAPS WEIGH US DOWN AND HOW WE RISE ABOVE THEM

244 *rat pups in gondolas:* Yoh'i Zennou-Azogui et al., "Hypergravity Within a Critical Period Impacts on the Maturation of Somatosensory Cortical Maps and Their Potential for Use-Dependent Plasticity in the Adult," *Journal of Neurophysiology* 115 (2016): 2740–60.

246 *right hemisphere stopped:* Lars Muckli et al., "Bilateral Visual Field Maps in a Patient with Only One Hemisphere," *Proceedings of the National Academy of Sciences of the United States of America* 106 (2009): 13034–39.

247 *could redirect messages:* Mriganka Sur et al., "Experimentally Induced Visual Projections into Auditory Thalamus and Cortex," *Science* 242 (1988): 1437–41.

map of visual space: Anna W. Roe et al., "A Map of Visual Space Induced in Primary Auditory Cortex," *Science* 250 (1990): 818–20.

seeing rather than hearing: Laurie von Melchner et al., "Visual Behaviour Mediated by Retinal Projections Directed to the Auditory Pathway," *Nature* 404 (2000): 871–76.

if they suffer a stroke: Angela O. Ballantyne et al., "Plasticity in the Developing Brain: Intellectual, Language, and Academic Functions in Children with Ischaemic Perinatal Stroke," *Brain* 131 (2008): 2975–85.

250 *employ an analogy:* George Lakoff & Mark Johnson, *Metaphors We Live By* (Chicago: University of Chicago Press, 2003).

251 *rates of metabolic disruptions:* Tessa Roseboom et al., "The Dutch Famine and Its Long-Term Consequences for Adult Health," *Early Human Development* 82 (2006): 485–91.

licking their pups: Josie Diorio & Michael Meaney, "Maternal Programming of Defensive Responses Through Sustained Effects on Gene Expression," *Journal of Psychiatry & Neuroscience* 32 (2007): 275–84.

252 *reactions to potential threats:* Willem E. Frankenhuis & Carolina de Weerth, "Does Early-Life Exposure to Stress Shape or Impair Cognition?" *Current Directions in Psychological Science* 22 (2013): 407–12.

physical and mental illness: Christine Heim & Charles B. Nemeroff, "The Role of Childhood Trauma in the Neurobiology of Mood and

Anxiety Disorders: Preclinical and Clinical Studies," *Biological Psychiatry* 49 (2001): 1023–39.

biological embedding: Clyde Hertzman, "The Biological Embedding of Early Experience and Its Effects on Health in Adulthood," *Annals of the New York Academy of Sciences* 896 (1999): 85–95.

higher rates of illness: Clyde Hertzman, "Putting the Concept of Biological Embedding in Historical Perspective," *Proceedings of the National Academy of Sciences of the United States of America* 109 (2012): 17160–67.

253 *recorded at a baby's scalp:* Dennis L. Molfese, "Predicting Dyslexia at 8 Years of Age Using Neonatal Brain Responses," *Brain and Language* 72 (2000): 238–45.

Index

Page numbers in *italics* refer to illustrations.